Deep-Sky Name Index 2000.0

Hugh C. Maddocks

Foxon-Maddocks Associates

Published by Foxon-Maddocks Associates, 10807 Oldfield Drive, Reston, Virginia 22091–5207 U.S.A.

Design, layout, and computer typesetting by Nancy Foxon Adams, Foxon Enterprises, 10907 Knights Bridge Court, Reston, Virginia 22090–3932 U.S.A.

Library of Congress Catalog Card Number: 90-84968

ISBN 0–9628305–0–X

Printed in the United States of America

10 9 8 7 6 5 4 3 2

To Ebbe Curtis Hoff, M.D.,
my mentor who made stargazing something special

Contents

List of Tables

Preface

I enjoyed astronomy as a teenager in the fifties but had let it take a back seat to other things until the fall of '89. When I started reading about current interests in deep-sky objects, it was clear that clusters, galaxies, and nebulae had become very popular. Double stars were no longer the primary objects being observed.

Because many of the popular deep-sky objects have names in addition to various catalog designations, I felt overwhelmed when I tried to memorize this information in addition to relearning the common names and Bayer designations for bright stars as well as the Latin names, common names, and abbreviations for constellations. Trying to remember that the Andromeda Galaxy, M31, and NGC 224 are the same thing; that the Beehive Cluster, M44, the Manger, NGC 2632, and the Praesepe Cluster are the same thing; that the star beta (β) Canis Minoris is also called Gomeisa, Gomeiza, or Gomelza and on and on was just too much for me.

That's when the idea for *Deep-Sky Name Index 2000.0* developed. I thought that an index of deep-sky object names could pull scattered information together, especially if the indexed information included the type of object, constellation in which the object is located, synonyms if any, and accurate epoch 2000.0 coordinates for use with star atlases and telescopes. Also, I wanted an index that would instantly tell me which named objects are located in any particular constellation so that I could plan what objects to observe at different times of year. To make the retrieval of information on any object as fast as possible, I decided to list all names for objects both individually and according to constellation. What I first thought would be a pamphlet has evolved into this book.

The index is intended to be a comprehensive source for names, but it is not intended to be a duplication of catalogue designations. A complete Messier catalogue, however, is included as an integral part of the index. Also, named objects that do have NGC or IC numbers are listed both by names and by those numbers.

I realize that there can be different opinions as to what should be included, and I know that a few errors will remain regardless of the number of times that each line is proofread and cross-checked with different sources. Additions and corrections to the index are welcomed from both amateur and professional astronomers. Please use the feedback form in the back of the book for suggested changes.

Deep-Sky Name Index 2000.0 is the handy reference that I wish I'd had when I started to learn about deep-sky objects. I hope that it saves time spent inside that could be spent outside under the stars.

Hugh C. Maddocks
Reston, Virginia

October 9, 1990

Introduction

The index has been designed to be easy to use. It allows quick access to information on any particular named deep-sky object or all named deep-sky objects in any particular constellation. Every item listed in the index includes the type of object, constellation in which the object is located, synonyms if any, and epoch 2000.0 coordinates[1].

Categories Used

The categories used for each object in the index are: clusters, galaxies, nebulae, and stars. These categories are also used in subdividing objects within constellations.

Clusters are subdivided into globular clusters and open clusters. ("Open" clusters are the same as what some references call "galactic" clusters.) Nebulae are subdivided into bright nebulae (which include emission and reflection nebulae), dark nebulae, and planetary nebulae. Galaxies and stars are not subdivided into other categories.

Listings for Constellations

Constellations are listed according to their Latin names, three-letter abbreviations adopted by the International Astronomical Union (IAU), as well as any other common names.

---

[1] Epoch 2000.0 coordinates are a pair of numbers listed in the order *right ascension, declination. Right ascension* is the coordinate which is similar to longitude; it is given as hours (h) and minutes (m) around the celestial equator from the vernal equinox. *Declination* is the coordinate similar to latitude; it is given as degrees (°) and minutes (') north (+) or south (−) of the celestial equator. Since the earth's axis traces out a large circle in the sky once every 26,000 years, coordinates are exact at only one particular moment, but are usually given in a standard epoch (time). The International Astronomical Union (IAU) has adopted 2000.0 (the start of the year 2000) as the current standard epoch. From a practical point of view, coordinates specified for epoch 2000.0 are very accurate from the late 20th century through the early 21st century.

Introduction

When a constellation contains one or more named stellar or nonstellar objects, the listings of the abbreviation and any common names are followed by *See* cross-references which refer to the Latin name. That is, named objects are listed according to constellation once under the listing of the Latin name.

The Latin name listing always includes the three-letter abbreviation as well as any other common names. For example, the Latin name listing for Aquila begins with the line

> **Aquila** (abbreviated **Aql**, Eagle):

while the listings for the abbreviation and common name are the lines

> **Aql**. *See* **Aquila**
>
> Eagle. *See* **Aquila**

When a constellation has no named objects, then the Latin name, abbreviation, and any common names are each listed without *See* references. For example, the Southern Crown is listed four times as

> **Corona Australis** (abbreviated **CrA**, Southern Crown)
>
> **CrA** (abbreviation for **Corona Australis**, Southern Crown)
>
> Crowns:
>> Northern Crown. *See* **Corona Borealis**
>>
>> Southern Crown (**Corona Australis**, abbreviated **CrA**)
>
> Southern Crown (**Corona Australis**, abbreviated **CrA**)

Listings for Stars

Stars are listed in the index according to both all common names for the star and the Bayer designation[2] for the star, if any.

---

[2] Bayer designations for stars were introduced by the Bavarian lawyer and astronomy aficionado Johann Bayer (1572–1625) in the first edition of his star atlas of 1603. The system labels bright stars using lower-case Greek letters followed by the genitive (possessive case) form of the constellation's Latin name (for example, α Andromedae). For constellations with more than 24 bright stars, both lower-case and upper-case Roman letters are also used.

Tables 1 and 2 are included to provide quick references for use with Bayer designations. Table 1 lists the names of the lower-case Greek letters, and Table 2 gives the nominative form of each Latin constellation name corresponding to the genitive (possessive case) form used in the Bayer designation. Several stars having only designations beginning with Roman letters are also included in the index.

In addition to the main listing(s), each star is also listed under the category "stars" in the constellation in which it is located. If the star has a Bayer designation, it is listed under the constellation according to that designation followed by any other names.

For example, the brightest star in the constellation Bootes is listed three times as

alpha (α) Bootis (Arcturus), *14h 15.7m, +19° 11'*

Arcturus (alpha (α) Bootis), *14h 15.7m, +19° 11'*

Bootes (abbreviated **Boo**, Bear Driver, Herdsman):
    stars:
        alpha (α) Bootis (Arcturus), *14h 15.7m, +19° 11'*

(The names of Greek letters are not capitalized in the index as a reminder that Bayer designations use lower-case Greek letters.)

If a named star does not have a Bayer designation but does have a Flamsteed number[3], then the Flamsteed number is listed following the name(s). For example, the star Atlas in the constellation Taurus is listed as

Atlas (27 Tauri), *3h 49.2m, +24° 03'*

Taurus (abbreviated **Tau**, Bull):
    stars:
        Atlas (27 Tauri), *3h 49.2m, +24° 03'*

---

[3] Flamsteed numbers for bright stars consist of Arabic numbers followed by the genitive (possessive case) form of the constellation's Latin name (for example, 61 Cygni). Flamsteed numbers were *not* introduced by John Flamsteed (1646–1719), the first Astronomer Royal of England. Instead, the French astronomer Joseph Jerome de Lalande (1732–1807) assigned these numbers to stars in a French edition of Flamsteed's work published in the 1780's. The stars in Flamsteed's atlas are labeled with Bayer designations.

Table 1

The Greek Alphabet

Lower-case Letter	Name
α	alpha
β	beta
γ	gamma
δ	delta
ε	epsilon
ζ	zeta
η	eta
θ	theta
ι	iota
κ	kappa
λ	lambda
μ	mu
ν	nu
ξ	xi
ο	omicron
π	pi
ρ	rho
σ	sigma
τ	tau
υ	upsilon
φ	phi
χ	chi
ψ	psi
ω	omega

Table 2

Constellation Names

<u>Genitive Form</u>	<u>Nominative Form</u>
Andromedae	Andromeda
Antliae	Antlia
Apodis	Apus
Aquarii	Aquarius
Aquilae	Aquila
Arae	Ara
Arietis	Aries
Aurigae	Auriga
Bootis	Bootes
Caeli	Caelum
Camelopardalis	Camelopardalis
Cancri	Cancer
Canis Majoris	Canis Major
Canis Minoris	Canis Minor
Canum Venaticorum	Canes Venatici
Capricorni	Capricornus
Carinae	Carina
Cassiopeiae	Cassiopeia
Centauri	Centaurus
Cephei	Cepheus
Ceti	Cetus
Chamaeleonis	Chamaeleon
Circini	Circinus
Columbae	Columba
Comae Berenices	Coma Berenices
Coronae Australis	Corona Australis
Coronae Borealis	Corona Borealis
Corvi	Corvus

Table 2 (continued)

<u>Genitive Form</u>	<u>Nominative Form</u>
Crateris	Crater
Crucis	Crux
Cygni	Cygnus
Delphini	Delphinus
Doradus	Dorado
Draconis	Draco
Equulei	Equuleus
Eridani	Eridanus
Fornacis	Fornax
Geminorum	Gemini
Gruis	Grus
Herculis	Hercules
Horologii	Horologium
Hydrae	Hydra
Hydri	Hydrus
Indi	Indus
Lacertae	Lacerta
Leonis	Leo
Leonis Minoris	Leo Minor
Leporis	Lepus
Librae	Libra
Lupi	Lupus
Lyncis	Lynx
Lyrae	Lyra
Mensae	Mensa
Microscopii	Microscopium
Monocerotis	Monoceros
Muscae	Musca
Normae	Norma
Octantis	Octans
Ophiuchi	Ophiuchus
Orionis	Orion

Table 2 (concluded)

<u>Genitive Form</u>	<u>Nominative Form</u>
Pavonis	Pavo
Pegasi	Pegasus
Persei	Perseus
Phoenicis	Phoenix
Pictoris	Pictor
Piscis Austrini	Piscis Austrinus
Piscium	Pisces
Puppis	Puppis
Pyxidis	Pyxis
Reticuli	Reticulum
Sagittae	Sagitta
Sagittarii	Sagittarius
Scorpii	Scorpius
Sculptoris	Sculptor
Scuti	Scutum
Serpentis	Serpens
Sextantis	Sextans
Tauri	Taurus
Telescopii	Telescopium
Trianguli	Triangulum
Trianguli Australis	Triangulum Australe
Tucanae	Tucana
Ursae Majoris	Ursa Major
Ursae Minoris	Ursa Minor
Velorum	Vela
Virginis	Virgo
Volantis	Volans
Vulpeculae	Vulpecula

Introduction

Most of the common names for bright stars are Arabic names which have been transliterated into English. Since there is no universally accepted form for the transliteration of Arabic, many of the Arabic star names can be spelled in more than one way. When there are multiple spellings of a name, the star is listed according to each of the possible spellings. Thus, for example, the brightest star in the constellation Aquila is listed four times as

> alpha (α) Aquilae (Altair, Atair), *19h 50.8m, +8° 52'*
>
> Altair (alpha (α) Aquilae, Atair), *19h 50.8m, +8° 52'*
>
> Atair (alpha (α) Aquilae, Altair), *19h 50.8m, +8° 52'*
>
> **Aquila** (abbreviated **Aql**, Eagle):
>> stars:
>>> alpha (α) Aquilae (Altair, Atair), *19h 50.8m, +8° 52'*

Some judgment had to exercised in deciding how many variations of transliterated Arabic names should be included. The index is intended to be reasonably comprehensive but could not be allowed to grow to the size of a Manhattan telephone book. If a particular name variation is not listed, the easy way to find the name(s) of the star is to look at the listing according to the Bayer designation, as in the example

> alpha (α) Aquilae (Altair, Atair), *19h 50.8m, +8° 52'*

Superscripts are used in star designations to indicate a member of a double or multiple star. Thus, the two main components of the famous Double-Double star in Lyra are listed in the index as epsilon[1] and epsilon[2] (ε^1 and ε^2) Lyrae.

Listings for Nonstellar Objects

Nonstellar objects (clusters, nebulae, and galaxies) are listed in the index according to all common names for the object, the Messier number[4] (if any), and the NGC or IC number[5] (if any). To

---

[4] Messier numbers were assigned to objects by the French astronomer Charles Messier (1730–1817), whose main purpose was to prevent the objects from being mistaken for comets. The index listings include M1 through M110, although only objects through M103 appear in Messier's original work. M104 through M109 were added by Messier's contemporary Pierre Méchain (1744–1805). There is evidence that Messier knew of

provide a complete Messier catalogue as part of the index, all Messier objects are listed by both their Messier numbers and NGC or IC numbers even though many of the Messier objects do not have names other than their Messier numbers.

In addition to the main listing(s), each nonstellar object is also listed under one of the following six categories in the constellation in which it is located

    clusters, globular
    clusters, open
    galaxies
    nebulae, bright
    nebulae, dark
    nebulae, planetary

If the object has a Messier number, then it is listed under the constellation according to that number followed by other names if any and NGC or IC designation if any.

For example, the Pleiades open cluster is listed four times as

M45 open cluster in Taurus (Pleiades, Seven Sisters},
    3h 47.0m, +24°07'

Pleiades open cluster in Taurus (M45, Seven Sisters),
    3h 47.0m, +24°07'

Seven Sisters open cluster in Taurus (M45, Pleiades),
    3h 47.0m, +24°07'

Taurus (abbreviated **Tau**, Bull):
    clusters, open:
        M45 (Pleiades, Seven Sisters), *3h 47.0m, +24°07'*

(Individual stars in the Pleiades are also listed both according to their names and in the listing for Taurus under the category "stars.")

---

the galaxy NGC 205 in Andromeda, which has been proposed as M110.

[5] NGC and IC numbers were assigned by the Danish astronomer Johann Louis Emil (J.L.E.) Dreyer (1852–1926), who compiled observations on a total of 13,226 objects, mostly nonstellar. The observations were published as the classic *New General Catalogue of Nebulae and Clusters of Stars* in 1888 (the NGC), the *Index Catalogue of 1895*, and the *Second Index Catalogue of 1908* (the ICs).

References

The following references were used to compile *Deep-Sky Name Index 2000.0.*

Allen, Richard Hinckley. 1963. *Star Names, Their Lore and Meaning.* New York: Dover. The Dover edition is an unabridged and corrected republication of the work first published by G. E. Stechert in 1899 under the former title: *Star-Names and Their Meanings.*

Astronomy magazine. October 1989 through September 1990 (Vol. 17 No. 10 through Vol. 18 No. 9) monthly star charts for the northern sky. Waukesha, Wis.: Kalmbach Publishing Co.

Burnham, Robert, Jr. 1978. Vol. 1: Andromeda—Cetus, Vol. 2, Chamaeleon—Orion, and Vol. 3: Pavo—Vulpecula. *Burnham's Celestial Handbook.* New York: Dover. The Dover edition is an expanded and updated republication of the work originally published by Celestial Handbook Publications, Flagstaff, Arizona, in 1966.

Hirshfield, Alan and Sinnott, Roger W., eds. 1982 (Vol. 1: Stars to Magnitude 8.0) and 1985 (Vol. 2: Double Stars, Variable Stars and Nonstellar Objects). *Sky Catalogue 2000.0.* Cambridge, Mass.: Sky Publishing Corp. and New York: Cambridge University Press.

Jones, Kenneth Glyn, ed. 1986 (Vol. 1: Double Stars. 2nd ed.), 1979 (Vol. 2: Planetary and Gaseous Nebulae), 1980 (Vol. 3: Open and Globular Cluster), 1981 (Vol. 4: Galaxies), 1982 (Vol. 5: Clusters of Galaxies), 1987 (Vol. 6: Anonymous Galaxies), and 1987 (Vol. 7: The Southern Sky). *Webb Society Deep-Sky Observer's Handbook.* Hillside, N.J.: Enslow Publishers.

Mallas, John H. and Kreimer, Evered. 1978. *The Messier Album.* Cambridge, Mass.: Sky Publishing Corp.

References

Mayall, R. Newton and Mayall, Margaret W. 1954. *Olcott's Field Book of the Skies*. 4th ed. New York: G. P. Putnam's Sons.

Menzel, Donald H. and Pasachoff, Jay M. 1983. *A Field Guide to Stars and Planets*. 2nd ed. Boston: Houghton Mifflin.

Muirden, James. 1987. *The Amateur Astronomer's Handbook*. 3rd ed. New York: Harper & Row.

Neely, Henry M. 1946. *A Primer for Star-Gazers*. New York: Harper & Brothers.

Sinnott, Roger W., ed. 1988. *NGC 2000.0*. Cambridge, Mass.: Sky Publishing Corp. and New York: Cambridge University Press.

Sky & Telescope magazine. October 1989 through September 1990 (Vol. 78 No. 4 through Vol. 80 No. 3) monthly star charts for the northern sky and bimonthly star charts for the southern sky. Cambridge, Mass.: Sky Publishing Corp.

Tirion, Wil. 1981 (reprinted in 1989). *Sky Atlas 2000.0*. Cambridge, Mass.: Sky Publishing Corp. and New York: Cambridge University Press.

Tirion, Wil, Rappaport, Barry, and Lovi, George. 1987. Vol. I: The Northern Hemisphere to -6° and Vol. II: The Southern Hemisphere to +6° (both reprinted in 1989). *Uranometria 2000.0*. Richmond, Va.: Willmann-Bell, Inc.

Deep-Sky Name Index 2000.0

In the index, **boldface** is used for primary listings of Latin names and three-letter IAU abbreviations for constellations. *Italics* is used for epoch 2000.0 coordinates. Items beginning with Greek letters are filed according to the name of the letter, and items beginning with numbers are filed as though the number were spelled out. Superscripts in star designations indicate a member of a double or multiple star.

a Carinae, *9h 11.0m, –58°58'*

Abt's Star in Crater, *11h 17.0m, –7°08'*

Acamar (theta¹ (θ¹) Eridani), *2h 58.3m, –40°18'*

Achernar (alpha (α) Eridani), *1h 37.7m, –57°14'*

Acrab (Akrab, beta¹ (β¹) Scorpii, Graffias), *16h 05.4m, –19°48'*

Acrux (alpha¹ (α¹) Crucis), *12h 26.6m, –63°06'*

Acubens (alpha (α) Cancri), *8h 58.5m, +11°51'*

Adara (Adhara, epsilon (ε) Canis Majoris), *6h 58.6m, –28°58'*

Adhafera (Aldhafera, zeta (ζ) Leonis), *10h 16.7m, +23°25'*

Adhara (Adara, epsilon (ε) Canis Majoris), *6h 58.6m, –28°58'*

Agena (beta (β) Centauri, Hadar, Wazn), *14h 03.8m, –60°22'*

Ain (epsilon (ε) Tauri), *4h 28.6m, +19°11'*

Air Pump, Pump (**Antlia**, abbreviated **Ant**)

Akrab (Acrab, beta¹ (β¹) Scorpii, Graffias), *16h 05.4m, –19°48'*

Albali (epsilon (ε) Aquarii), *20h 47.7m, –9°30'*

Albireo (beta (β) Cygni), *19h 30.7m, +27°58'*

Alcaid (Alkaid, Benetnasch, eta (η) Ursae Majoris), *13h 47.5m, +49°19'*

Alchemb (Algenib, alpha (α) Persei, Marfak, Marfik, Mirfak), *3h 24.3m, +49°52'*

Alchiba (alpha (α) Corvi), *12h 08.4m, –24°44'*

Alcor (80 Ursae Majoris, g Ursae Majoris, Saidak), *13h 25.2m, +54°59'*

Alcyone (eta (η) Tauri), *3h 47.5m, +24°06'*

Alcyone (bright) Nebula in Taurus, *3h 47.5m, +24°06'*

Aldebaran (alpha (α) Tauri, Cor Tauri, Palilicium), *4h 35.9m, +16°31'*

Alderamin (alpha (α) Cephei), *21h 18.6m, +62°35'*

Aldhafera (Adhafera, zeta (ζ) Leonis), *10h 16.7m, +23°25'*

Alfard (alpha (α) Hydrae, Alphard, Cor Hydrae, Red Bird),
9h 27.6m, –8° 40'

Alferats (alpha (α) Andromedae, Alpheratz, Sirrah), *0h 08.4m,
+29° 05'*

Alfeta (alpha (α) Coronae Borealis, Alphecca, Gemma), *15h 34.7m,
+26° 43'*

Alfirk (Alphirk, beta (β) Cephei), *21h 28.7m, +70° 34'*

Algebar (beta (β) Orionis, Elgebar, Rigel), *5h 14.5m, –8° 12'*

Algedi (Algiedi, alpha² (α²) Capricorni, Giedi), *20h 18.1m, –12° 33'*

Algeiba (Algieba, gamma¹ (γ¹) Leonis), *10h 20.0m, +19° 50'*

Algenib:

    alpha (α) Persei (Alchemb, Marfak, Marfik, Mirfak), *3h 24.3m,
      +49° 52'*

    gamma (γ) Pegasi, *0h 13.2m, +15° 11'*

Algenubi (epsilon (ε) Leonis), *9h 45.8m, +23° 46'*

Algieba (Algeiba, gamma¹ (γ¹) Leonis), *10h 20.0m, +19° 50'*

Algiedi (Algedi, alpha² (α²) Capricorni, Giedi), *20h 18.1m, –12° 33'*

Algol (beta (β) Persei, Demon Star, Medusa Star), *3h 08.2m,
+40° 57'*

Algorab (Algores, delta (δ) Corvi), *12h 29.9m, –16° 31'*

Alhena (Almeisam, gamma (γ) Geminorum), *6h 37.7m, +16° 24'*

Alioth (epsilon (ε) Ursae Majoris), *12h 54.0m, +55° 58'*

Alkaid (Alcaid, Benetnasch, eta (η) Ursae Majoris), *13h 47.5m,
+49° 19'*

Alkalurops (mu¹ (μ¹) Bootis), *15h 24.5m, +37° 23'*

Alkes (alpha (α) Crateris), *10h 59.8m, –18° 18'*

Almach (Almak, gamma¹ (γ¹) Andromedae), *2h 03.9m, +42° 20'*

Almeisam (Alhena, gamma (γ) Geminorum), *6h 37.7m, +16° 24'*

Alnair (alpha (α) Gruis), *22h 08.2m, –46° 58'*

Alnasl (gamma (γ) Sagittarii), *18h 05.8m, –30° 25'*

Alnath (beta (β) Tauri, Elnath, Nath), *5h 26.3m, +28° 36'*

Alnilam (Anilam, epsilon (ε) Orionis), *5h 36.2m, –1° 12'*

Alnitak (zeta (ζ) Orionis), *5h 40.8m, –1° 57'*

Alniyat (sigma (σ) Scorpii), *16h 21.2m, –25° 36'*

alpha (α) Andromedae (Alferats, Alpheratz, Sirrah), *0h 08.4m, +29° 05'*

alpha (α) Andromedae and alpha, beta, & gamma (α, β, & γ) Pegasi (Great Square of Pegasus), *0h 08.4m, +29° 05' and 23h 04.8m, +15° 12'; 23h 03.8m, +28° 05'; & 0h 13.2m, +15° 11'*

alpha (α) Aquarii (Sadalmelik), *22h 05.8m, –0° 19'*

alpha (α) Aquilae (Altair, Atair), *19h 50.8m, +8° 52'*

alpha (α) Aquilae, alpha (α) Cygni, and alpha (α) Lyrae (Summer Triangle), *19h 50.8m, +8° 52' ; 20h 41.4m, +45° 17'; and 18h 36.9m, +38° 47'*

alpha (α) Arietis (Hamal), *2h 07.2m, +23° 28'*

alpha (α) Aurigae (Capella, Goat Star), *5h 16.7m, +46° 00'*

alpha (α) Bootis (Arcturus), *14h 15.7m, +19° 11'*

alpha (α) Cancri (Acubens), *8h 58.5m, +11° 51'*

alpha (α) Canis Majoris (Dog Star, Nile Star, Sirius), *6h 45.1m, –16° 43'*

alpha (α) Canis Minoris (Little Dog Star, Procyon), *7h 39.3m, +5° 14'*

alpha² (α²) Canum Venaticorum (Cor Caroli), *12h 56.0m, +38° 19'*

alpha² (α²) Capricorni (Algedi, Algiedi, Giedi), *20h 18.1m, –12° 33'*

alpha (α) Carinae (Canopus), *6h 24.0m, –52° 42'*

alpha (α) Cassiopeiae (Schedar, Schedir, Shedar), *0h 40.5m, +56° 32'*

alpha¹ (α¹) Centauri (Rigil Kentaurus, Toliman), *14h 39.6m, –60° 50'*

alpha (α) Cephei (Alderamin), *21h 18.6m, +62° 35'*

alpha (α) Ceti (Menkab, Menkar), *3h 02.3m, +4° 05'*

alpha (α) Columbae (Phact, Phaet), *5h 39.6m, –34° 04'*

alpha (α) Coronae Borealis (Alfeta, Alphecca, Gemma), *15h 34.7m, +26° 43'*

alpha (α) Corvi (Alchiba), *12h 08.4m, –24° 44'*

alpha (α) Crateris (Alkes), *10h 59.8m, –18° 18'*

alpha¹ (α¹) Crucis (Acrux), *12h 26.6m, –63° 06'*

alpha (α) Cygni (Arided, Deneb), *20h 41.4m, +45° 17'*

alpha (α) Cygni, alpha (α) Aquilae, and alpha (α) Lyrae (Summer Triangle), *20h 41.4m, +45° 17'; 19h 50.8m, +8° 52' ; and 18h 36.9m, +38° 47'*

alpha (α) Delphini (Sualocin, Svalocin), *20h 39.6m, +15° 55'*

alpha (α) Draconis (Thuban), *14h 04.4m, +64° 23'*

alpha (α) Equulei (Kitalpha), *21h 15.8m, +5° 15'*

alpha (α) Eridani (Achernar), *1h 37.7m, –57° 14'*

alpha (α) Geminorum (Castor), *7h 34.6m, +31° 53'*

alpha (α) Gruis (Alnair), *22h 08.2m, –46° 58'*

alpha¹ (α¹) Herculis (Rasalgethi), *17h 14.6m, +14° 23'*

alpha (α) Hydrae (Alfard, Alphard, Cor Hydrae, Red Bird), *9h 27.6m, –8° 40'*

alpha (α) Leonis (Cor Leonis, Regulus), *10h 08.4m, +11° 58'*

alpha (α) Leporis (Arneb, Arsh), *5h 32.7m, –17° 49'*

alpha² (α²) Librae (Kiffa Australis, Zubenelgenubi), *14h 50.9m, –16° 02'*

alpha (α) Lyrae (Harp Star, Vega, Wega), *18h 36.9m, +38° 47'*

alpha (α) Lyrae, alpha (α) Aquilae, and alpha (α) Cygni (Summer Triangle), *18h 36.9m, +38° 47'; 19h 50.8m, +8° 52' ; and 20h 41.4m, +45° 17'*

alpha (α) Ophiuchi (Rasalague, Rasalhague), *17h 34.9m, +12° 34'*

alpha (α) Orionis (Betelgeuse, Martial Star, Mirzam), *5h 55.2m, +7° 24'*

alpha (α) Pavonis (Peacock Star), *20h 25.6m, –56° 44'*

alpha (α) Pegasi (Marchab, Markab), *23h 04.8m, +15° 12'*

alpha, beta, & gamma (α, β, & γ) Pegasi and alpha (α) Andromedae (Great Square of Pegasus), *23h 04.8m, +15° 12'; 23h 03.8m, +28° 05'; & 0h 13.2m, +15° 11'; and 0h 08.4m, +29° 05'*

alpha (α) Persei (Alchemb, Algenib, Marfak, Marfik, Mirfak), *3h 24.3m, +49° 52'*

alpha (α) Persei (open) Cluster, *3h 22.0m, +49° 00'*

alpha (α) Phoenicis (Ankaa), *0h 26.3m, –42° 18'*

alpha (α) Piscis Austrini (First Frog, Fomalhaut), *22h 57.6m, –29° 37'* alpha (α) Piscium (Alrescha, Alrischa, Kaitain, Okda), *2h 02.0m, +2° 46'*

alpha (α) Sagittarii (Rukbat), *19h 23.9m, –40° 37'*

alpha (α) Scorpii (Antares, Cor Scorpii), *16h 29.4m, –26° 26'*

alpha (α) Serpentis (Cor Serpentis, Unukalhai), *15h 44.3m, +6° 26'*

alpha (α) Tauri (Aldebaran, Cor Tauri, Palilicium), *4h 35.9m, +16° 31'*

alpha (α) Trianguli (Caput Trianguli, Rasalmuthallath), *1h 53.1m, +29° 35'*

alpha (α) Trianguli Australis (Atria), *16h 48.7m, –69° 02'*

alpha (α) Ursae Majoris (Dubhe), *11h 03.7m, +61° 45'*

alpha and beta (α and β) Ursae Majoris (Pointer Stars), *11h 03.7m, +61° 45' and 11h 01.8m, +56° 23'*

alpha (α) Ursae Minoris (Cynosura, North Star, Polaris, Pole Star), *2h 31.8m, +89° 16'*

alpha (α) Virginis (Azimech, Spica), *13h 25.2m, –11° 10'*

Alphard (Alfard, alpha (α) Hydrae, Cor Hydrae, Red Bird), *9h 27.6m, –8° 40'*

Alphecca (Alfeta, alpha (α) Coronae Borealis, Gemma), *15h 34.7m, +26° 43'*

Alpheratz (Alferats, alpha (α) Andromedae, Sirrah), *0h 08.4m, +29° 05'*

Alphirk (Alfirk, beta (β) Cephei), *21h 28.7m, +70° 34'*

Alrai (Errai, gamma (γ) Cephei), *23h 39.3m, +77° 38'*

Alrakis (Arrakis, mu (μ) Draconis), *17h 05.3m, +54° 28'*

Alrescha (alpha (α) Piscium, Alrischa, Kaitain, Okda), *2h 02.0m, +2° 46'*

Alshain (beta (β) Aquilae), *19h 55.3m, +6° 24'*

Alshemali (mu (μ) Leonis, Rasalas), *9h 52.8m, +26° 00'*

Altair (alpha (α) Aquilae, Atair), *19h 50.8m, +8° 52'*

Altais (delta (δ) Draconis, Nodus Secundus), *19h 12.6m, +67° 40'*

Altar (**Ara**, abbreviated **Ara**)

Alterf (lambda (λ) Leonis), *9h 31.7m, +22° 58'*

Aludra (eta (η) Canis Majoris), *7h 24.1m, –29° 18'*

Alula Australis (xi (ξ) Ursae Majoris, El Acola), *11h 18.2m,
+31° 32'*

Alula Borealis (nu (ν) Ursae Majoris), *11h 18.5m, +33° 06'*

Alwaid (beta (β) Draconis, Rastaban), *17h 30.4m, +52° 18'*

Alya (theta¹ (θ¹) Serpentis), *18h 56.2m, +4° 12'*

Amazon Star (Bellatrix, gamma (γ) Orionis), *5h 25.1m, +6° 21'*

Ambartsumian's Knot galaxy in Ursa Major (NGC 3561C),
11h 11.6m, +28° 43'

America (bright) Nebula in Cygnus (NGC 7000, North America
Nebula), *20h 58.8m, +44° 20'*

Ancha (theta (θ) Aquarii), *22h 16.8m, –7° 47'*

And. *See* **Andromeda**

Andrew's Star in Auriga, *5h 40.6m, +31° 22'*

Andromeda (abbreviated **And**, Chained Lady, Woman Chained):

galaxies:

Andromeda Galaxy (Great Galaxy in Andromeda, M31,
NGC 224), *0h 42.7m, +41° 16'*

Andromeda I (satellite of Andromeda Galaxy), *0h 45.7m,
+38° 00'*

Andromeda II (satellite of Andromeda Galaxy), *1h 16.4m,
+33° 27'*

Andromeda III (satellite of Andromeda Galaxy), *0h 35.4m,
+36° 31'*

Andromeda IV (satellite of Andromeda Galaxy), *0h 42.5m,
+40° 34'*

Baade A and B, *0h 49.9m, +42° 35'*

M31 (Andromeda Galaxy, Great Galaxy in Andromeda,
NGC 224), *0h 42.7m, +41° 16'*

M32 (NGC 221), *0h 42.7m, +40° 52'*

M110 (NGC 205), *0h 40.4m, +41° 41'*

nebula, planetary, Blue Snowball Nebula (NGC 7662),
23h 25.9m, +42° 33'

stars:

alpha (α) Andromedae (Alferats, Alpheratz, Sirrah),
0h 08.4m, +29° 05'

Andromeda (cont.):

    stars (cont.):

        alpha (α) Andromedae and alpha, beta, & gamma (α, β, & γ) Pegasi (Great Square of Pegasus), *0h 08.4m, +29°05′ and 23h 04.8m, +15°12′; 23h 03.8m, +28°05′; & 0h 13.2m, +15°11′*

        beta (β) Andromedae (Mirach), *1h 09.7m, +35°37′*

        gamma^1 (γ^1) Andromedae (Almach, Almak), *2h 03.9m, +42°20′*

        Lamont's Star, *0h 45.4m, +41°32′*

Andromeda Galaxy (Great Galaxy in Andromeda, M31, NGC 224), *0h 42.7m, +41°16′*

Andromeda I galaxy (satellite of Andromeda Galaxy), *0h 45.7m, +38°00′*

Andromeda II galaxy (satellite of Andromeda Galaxy), *1h 16.4m, +33°27′*

Andromeda III galaxy (satellite of Andromeda Galaxy), *0h 35.4m, +36°31′*

Andromeda IV galaxy (satellite of Andromeda Galaxy), *0h 42.5m, +40°34′*

Anilam (Alnilam, epsilon (ε) Orionis), *5h 36.2m, –1°12′*

Ankaa (alpha (α) Phoenicis), *0h 26.3m, –42°18′*

Ant (abbreviation for **Antlia**, Air Pump, Pump)

Ant (bright) Nebula in Norma, *16h 17.2m, –51°59′*

Antares (alpha (α) Scorpii, Cor Scorpii), *16h 29.4m, –26°26′*

Antares (bright) Nebula in Scorpius (Cloud Nebula), *16h 29.2m, –26°27′*

Antennae galaxies in Corvus (NGC 4038 and 4039, Ring-Tail Galaxy, Snorter), *12h 01.9m, –18°52′ and 12h 01.9m, –18°53′*

Antevorta (gamma (γ) Virginis, Porrima), *12h 41.7m, –1°27′*

Antlia (abbreviated **Ant**, Air Pump, Pump)

AO Cassiopeiae (Pearce's Star), *0h 17.7m, +51°26′*

Apus, (abbreviated **Aps**, Bird of Paradise)

Aql. *See* **Aquila**

Aquarius (abbreviated **Aqr**, Water Bearer, Water Jar):

    clusters, globular:

        M2 (NGC 7089), *21h 33.5m, –0° 49'*

        M72 (NGC 6981), *20h 53.5m, –12° 32'*

    clusters, open, M73 (NGC 6994), *20h 59.0m, –12° 38'*

    galaxy, Aquarius Dwarf Galaxy, *20h 46.9m, –12° 51'*

    nebulae, bright, Baxendell's Unphotographable Nebula (NGC 7088), *21h 33.4m, –0° 23'*

    nebulae, planetary:

        Helical Nebula (Helix Nebula, NGC 7293), *22h 29.6m, –20° 48'*

        Saturn Nebula (NGC 7009), *21h 04.2m, –11° 22'*

    stars:

        alpha (α) Aquarii (Sadalmelik), *22h 05.8m, –0° 19'*

        beta (β) Aquarii (Sadalsuud), *21h 31.6m, –5° 34'*

        Chevremont's Star, *21h 33.5m, –0° 49'*

        delta (δ) Aquarii (Scheat, Skat), *22h 54.6m, –15° 49'*

        epsilon (ε) Aquarii (Albali), *20h 47.7m, –9° 30'*

        gamma (γ) Aquarii (Sadachbia), *22h 21.7m, –1° 23'*

        kappa (κ) Aquarii (Situla), *22h 37.8m, –4° 14'*

        theta (θ) Aquarii (Ancha), *22h 16.8m, –7° 47'*

Aquarius Dwarf Galaxy, *20h 46.9m, –12° 51'*

Aquila (abbreviated **Aql**, Eagle):

    stars:

        alpha (α) Aquilae (Altair, Atair), *19h 50.8m, +8° 52'*

        alpha (α) Aquilae, alpha (α) Cygni, and alpha (α) Lyrae (Summer Triangle), *19h 50.8m, +8° 52'; 20h 41.4m, +45° 17'; and 18h 36.9m, +38° 47'*

        beta (β) Aquilae (Alshain), *19h 55.3m, +6° 24'*

        epsilon (ε) Aquilae (Deneb), *18h 59.6m, +15° 04'*

        f Aquilae (26 Aquilae), *19h 20.5m, –5° 25'*

        gamma (γ) Aquilae (Tarazed), *19h 46.3m, +10° 37'*

Aquila (cont.):

    stars (cont.):

        Luyten's Star, *19h 42.1m, +6° 00'*

        Sanduleak-Stephenson star (SS 433), *19h 11.8m, +4° 59'*

        Van Biesbroeck's Star, *19h 17.0m, +5° 09'*

        zeta (ζ) Aquilae (Deneb), *19h 05.4m, +13° 52'*

Ara (abbreviated **Ara**, Altar)

Archer. *See* **Sagittarius**

Arcturus (alpha (α) Bootis), *14h 15.7m, +19° 11'*

Argo Navis. *See* **Carina**; **Puppis**; **Pyxis**; **Vela**

Ari. *See* **Aries**

Arided (alpha (α) Cygni, Deneb), *20h 41.4m, +45° 17'*

Aries (abbreviated **Ari**, Ram):

    stars:

        alpha (α) Arietis (Hamal), *2h 07.2m, +23° 28'*

        beta (β) Arietis (Sheratan), *1h 54.6m, +20° 48'*

        gamma^2 (γ^2) Arietis (Mesartim), *1h 53.5m, +19° 18'*

Arkab (beta^1 (β^1) Sagittarii), *19h 22.6m, −44° 28'*

Arneb (alpha (α) Leporis, Arsh), *5h 32.7m, −17° 49'*

Arp's Galaxy in Ursa Major, *11h 19.6m, +51° 30'*

Arrakis, (Alrakis, mu (μ) Draconis), *17h 05.3m, +54° 28'*

Arrow. *See* **Sagitta**

Arsh (alpha (α) Leporis, Arneb), *5h 32.7m, −17° 49'*

Ascella (zeta (ζ) Sagittarii), *19h 02.6m, −29° 53'*

Asellus (theta (θ) Bootis), *14h 25.2m, +51° 51'*

Asellus Australis (delta (δ) Cancri), *8h 44.7m, +18° 09'*

Asellus Borealis (gamma (γ) Cancri), *8h 43.3m, +21° 28'*

Aspidiske (iota (ι) Carinae), *9h 17.1m, −59° 17'*

Asterope (Sterope, 21 Tauri), *3h 45.9m, +24° 33'*

Atair (alpha (α) Aquilae, Altair), *19h 50.8m, +8° 52'*

Atik (zeta (ζ) Persei), *3h 54.1m, +31° 53'*

Atlas (27 Tauri), *3h 49.2m, +24° 03'*

Atlas (bright) Nebula in Taurus, *3h 49.2m, +24° 03'*

Atria (alpha (α) Trianguli Australis), *16h 48.7m, –69° 02'*

Auriga (abbreviated **Aur**, Charioteer, Kite, Wagoner):

 clusters, open:

  M36 (NGC 1960), *5h 36.1m, +34° 08'*

  M37 (NGC 2099), *5h 52.4m, +32° 33'*

  M38 (NGC 1912), *5h 28.7m, +35° 50'*

 nebula, bright, Flaming Star Nebula (IC 405, Schaeberle's Flaming Star), *5h 16.2m, +34° 16'*

 stars:

  alpha (α) Aurigae (Capella, Goat Star), *5h 16.7m, +46° 00'*

  Andrew's Star, *5h 40.6m, +31° 22'*

  beta (β) Aurigae (Menkalinan), *5h 59.5m, +44° 57'*

  epsilon, eta, and zeta (ε, η, and ζ) Aurigae (Kids), *5h 02.0m, +43° 49'; 5h 06.5m, +41° 14'; and 5h 02.5m, +41° 05'*

  zeta (ζ) Aurigae (Sadatoni), *5h 02.5m, +41° 05'*

Avior (epsilon (ε) Carinae), *8h 22.5m, –59° 31'*

Azelfafage (pi^1 (π^1) Cygni), *21h 42.1m, +51° 11'*

Azha (eta (η) Eridani), *2h 56.4m, –8° 54'*

Azimech (alpha (α) Virginis, Spica), *13h 25.2m, –11° 10'*

B Cassiopeiae (Tycho's Star), *0h 25.3m, +64° 09'*

b Persei, *4h 18.2m, +50° 18'*

Baade A and B galaxies in Andromeda, *0h 49.9m, +42° 35'*

Babcock's Magnetic Star in Lacerta, *22h 44.2m, +55° 35'*

Baham (Biham, theta (θ) Pegasi), *22h 10.2m, +6° 12'*

Balance. *See* **Libra**

Barbon's Galaxy in Pegasus, *23h 37.7m, +30° 08'*

Barnard's Galaxy in Sagittarius (NGC 6822), *19h 44.9m, −14° 48'*

Barnard's Loop bright nebula in Orion (Orion Loop), centered at
    5h 35.0m, −3° 00'

Barnard's S (dark) Nebula in Ophiuchus (Dark S Nebula, S
    Nebula, Snake Nebula), *17h 23.5m, −23° 38'*

Barnard's Star in Ophiuchus (Runaway Star), *17h 57.8m, +4° 42'*

Baten Kaitos (zeta (ζ) Ceti), *1h 51.5m, −10° 20'*

Baxendell's Unphotographable (bright) Nebula in Aquarius (NGC
    7088), *21h 33.4m, −0° 23'*

Bear Claw Galaxy in Lynx (Bear Paw Galaxy, NGC 2537),
    8h 13.2m, +46° 00'

Bear Driver. *See* **Bootes**

Bear Paw Galaxy in Lynx (Bear Claw Galaxy, NGC 2537),
    8h 13.2m, +46° 00'

Bears:

    Greater Bear. *See* **Ursa Major**

    Lesser Bear. *See* **Ursa Minor**

Becklin's Star in Orion, *5h 35.3m, −5° 23'*

Beehive (open) Cluster in Cancer (M44, Manger, NGC 2632,
    Praesepe Cluster), *8h 40.1m, +19° 59'*

Beid (omicron^1 (o^1) Eridani), *4h 11.9m, −6° 50'*

Bellatrix (Amazon Star, gamma (γ) Orionis), *5h 25.1m, +6° 21'*

Belt of Orion (delta, epsilon, and zeta (δ, ε, and ζ) Orionis),
     5h 32.0m. –0° 18'; 5h 36.2m, –1° 12'; and 5h 40.8m, –1° 57'

Benetnasch (Alcaid, Alkaid, eta (η) Ursae Majoris), *13h 47.5m,
     +49° 19'*

Berenice's Hair. *See* **Coma Berenices**

beta (β) Andromedae (Mirach), *1h 09.7m, +35° 37'*

beta (β) Aquarii (Sadalsuud), *21h 31.6m, –5° 34'*

beta (β) Aquilae (Alshain), *19h 55.3m, +6° 24'*

beta (β) Arietis (Sheratan), *1h 54.6m, +20° 48'*

beta (β) Aurigae (Menkalinan), *5h 59.5m, +44° 57'*

beta (β) Bootis (Nakkar, Nekkar), *15h 01.9m, +40° 23'*

beta (β) Canis Majoris (Mirzam, Murzim), *6h 22.7m, –17° 57'*

beta (β) Canis Minoris (Gomeisa, Gomeiza, Gomelza), *7h 27.2m,
     +8° 17'*

beta (β) Canum Venaticorum (Chara), *12h 33.7m, +41° 21'*

beta (β) Capricorni (Dabih), *20h 21.0m, –14° 47'*

beta (β) Carinae (Miaplacidus), *9h 13.2m, –69° 43'*

beta (β) Cassiopeiae (Caph), *0h 09.2m, +59° 09'*

beta (β) Centauri (Agena, Hadar, Wazn), *14h 03.8m, –60° 22'*

beta (β) Cephei (Alfirk, Alphirk), *21h 28.7m, +70° 34'*

beta (β) Ceti (Deneb Kaitos, Difda, Diphda, Second Frog),
     0h 43.6m, –17° 59'

beta (β) Columbae (Wazn, Wezn), *5h 51.0m, –35° 46'*

beta (β) Coronae Borealis (Nusakan), *15h 27.8m, +29° 06'*

beta (β) Crucis (Mimosa), *12h 47.7m, –59° 41'*

beta (β) Cygni (Albireo), *19h 30.7m, +27° 58'*

beta (β) Delphini (Rotanev), *20h 37.5m, +14° 36'*

beta (β) Draconis (Alwaid, Rastaban), *17h 30.4m, +52° 18'*

beta (β) Eridani (Cursa, Dhalim, Kursa), *5h 07.8m, –5° 05'*

beta (β) Geminorum (Pollux), *7h 45.3m, +28° 02'*

beta (β) Herculis (Kornephoros), *16h 30.2m, +21° 29'*

beta (β) Leonis (Dafira, Deneb, Denebola, Serpha), *11h 49.1m,
     +14° 34'*

beta (β) Leporis (Nihal), *5h 28.2m, –20° 46'*

beta (β) Librae (Kiffa Borealis, Zubeneschamali), *15h 17.0m, –9° 23'*

beta (β) Lyrae (Sheliak, Shelyak), *18h 50.1m, +33° 22'*

beta (β) Ophiuchi (Cebalrai, Cheleb), *17h 43.5m, +4° 34'*

beta (β) Orionis (Algebar, Elgebar, Rigel), *5h 14.5m, –8° 12'*

beta (β) Pegasi (Scheat), *23h 03.8m, +28° 05'*

beta (β) Persei (Algol, Demon Star, Medusa Star), *3h 08.2m, +40° 57'*

beta[1] (β[1]) Sagittarii (Arkab), *19h 22.6m, –44° 28'*

beta[1] (β[1]) Scorpii (Acrab, Akrab, Graffias), *16h 05.4m, –19° 48'*

beta (β) Tauri (Alnath, Elnath, Nath), *5h 26.3m, +28° 36'*

beta (β) Ursae Majoris (Merak, Mirak), *11h 01.8m, +56° 23'*

beta (β) Ursae Minoris (Kochab), *14h 50.7m, +74° 09'*

beta (β) Virginis (Zavijava), *11h 50.7m, +1° 46'*

Betelgeuse (alpha (α) Orionis, Martial Star, Mirzam), *5h 55.2m, +7° 24'*

Bidelman's Helium Variable Star in Centaurus, *14h 23.0m, –39° 31'*

Bidelman's Peculiar Star in Perseus, *4h 48.9m, +43° 17'*

Bidelman's Star in Bootes, *14h 31.9m, +18° 45'*

Big Bear. *See* **Ursa Major**

Big Dipper. *See* **Ursa Major**

Big Dog. *See* **Canis Major**

Biham (Baham, theta (θ) Pegasi), *22h 10.2m, +6° 12'*

Birds:

    Bird of Paradise. *See* **Apus**

    Crane. *See* **Grus**

    Crow. *See* **Corvus**

    Dove. *See* **Columba**

    Eagle. *See* **Aquila**

    Peacock. *See* **Pavo**

    Phoenix. *See* **Phoenix**

Birds (cont.):

    Raven. *See* **Corvus**

    Swan. *See* **Cygnus**

    Toucan. *See* **Tucana**

Black Hole dark nebula in Sagittarius, *18h 15.5m, –18° 11'*

Blackeye Galaxy in Coma Berenices (M64, NGC 4826), *12h 56.7m, +21° 41'*

Blaze Star (T Coronae Borealis), *15h 59.5m, +25° 55'*

Blinking Planetary Nebula in Cygnus (NGC 6826), *19h 44.8m, +50° 31'*

Blue (planetary) Nebula in Centaurus (NGC 3918), *11h 50.3m, –57° 11'*

Blue Snowball (planetary) Nebula in Andromeda (NGC 7662), *23h 25.9m, +42° 33'*

Bode's "Nebulae" galaxies in Ursa Major (M81 and M82, NGC 3031 and 3034), *9h 55.6m, +69° 04'* and *9h 55.8m, +69° 41'*

Bok's Valentine dark nebula in Vela, *8h 25.5m, –51° 01'*

Bond's Flare Star in Sagittarius, *23h 31.7m, –2° 45'*

Boo. *See* **Bootes**

Boomerang (bright) Nebula in Centaurus, *12h 44.8m, –54° 31'*

Bootes (abbreviated **Boo**, Bear Driver, Herdsman):

    galaxies, Bootes Cluster, *14h 33.0m, +31° 33'*

    stars:

        alpha (α) Bootis (Arcturus), *14h 15.7m, +19° 11'*

        beta (β) Bootis (Nakkar, Nekkar), *15h 01.9m, +40° 23'*

        Bidelman's Star, *14h 31.9m, +18° 45'*

        d Bootis (12 Bootis), *14h 10.4m, +25° 06'*

        epsilon (ε) Bootis (Izar, Mizar, Pulcherrima), *14h 45.0m, +27° 04'*

        eta (η) Bootis (Muphrid), *13h 54.7m, +18° 24'*

        gamma (γ) Bootis (Seginus), *14h 32.1m, +38° 19'*

        i Bootis (44 Bootis), *15h 03.8m, +47° 39'*

        mu^1 (μ^1) Bootis (Alkalurops), *15h 24.5m, +37° 23'*

        theta (θ) Bootis (Asellus), *14h 25.2m, +51° 51'*

Bootes Cluster of galaxies, *14h 33.0m, +31° 33'*

Borrelly's Star in Cetus, *0h 23.8m, –9° 28'*

Box galaxies in Coma Berenices (NGC 4169, 4173, 4174, and 4175), *12h 12.2m, +29° 10'; 12h 12.3m, +29° 11'; 12h 12.4m, +29° 08'; and 12h 12.5m, +29° 09'*

Box (planetary) Nebula in Ophiuchus (NGC 6309), *17h 14.1m, –12° 54'*

Bradley 3077 star in Cassiopeia, *23h 13.3m, +57° 10'*

Branchett's Object star in Scutum, *18h 46.9m, –4° 57'*

Brewer's Star in Monoceros, *6h 52.0m, –1° 39'*

Bridal Veil (bright) Nebula in Cygnus (Cirrus Nebula; Cygnus Loop; NGC 6960, 6992, and 6995; Veil Nebula), *20h 45.7m, +30° 43'; 20h 56.4m, +31° 43'; and 20h 57.1m, +31° 13'*

Brocchi's (open) Cluster in Vulpecula (Coat Hanger), *19h 25.4m, +20° 11'*

Bubble (bright) Nebula in Cassiopeia (NGC 7635), *23h 20.7m, +61° 12'*

Bug (planetary) Nebula in Scorpius (NGC 6302), *17h 13.7m, –37° 06'*

Bull. *See* **Taurus**

Burbidge Chain of galaxies in Cetus, *0h 47.5m, –20° 26'*

Burin. *See* **Caelum**

Burnham's (bright) Nebula in Taurus, *4h 22.0m, +19° 32'*

Butterfly (open) Cluster in Scorpius (M6, NGC 6405), *17h 40.1m, –32° 13'*

Butterfly (bright) Nebula in Carina (IC 2220, Toby Jug Nebula), *7h 56.8m, –59° 07'*

Butterfly (planetary) Nebula in Perseus (Cork Nebula, Little Dumbbell Nebula, M76, NGC 650 and 651), *1h 42.3m, +51° 34' and 1h 42.3m, +51° 35'*

BW Tauri Galaxy, *4h 33.2m, +5° 21'*

Cc

Caelum (abbreviated **Cae**, Burin, Chisel, Engraving Tool, Gravingtool):

    galaxies:

        Carafe Galaxy, *4h 28.0m, –47°54'*

        Carafe Group (NGC 1595, 1598, and Carafe Galaxy), *4h 28.4m, –47°49'; 4h 28.6m, –47°47'; and 4h 28.0m, –47°54'*

California (bright) Nebula in Perseus (NGC 1499), *4h 00.7m, +36°37'*

Camelopardalis (abbreviated **Cam**, Giraffe):

    cluster, open, Pazmino's Cluster, *3h 16.3m, +60°02'*

    galaxies:

        Holmberg III, *9h 14.8m, +74°14'*

        Integral Sign Galaxy, *7h 11.4m, +71°50'*

    nebula, dark, Thumbprint Nebula, *12h 42.9m, +78°16'*

    stars, Kemble's Cascade star chain, *3h 57.0m, +63°00'*

Camelopardus. *See* **Camelopardalis**

Campbell's Star in Cygnus, *19h 34.8m, +30°31'*

Cancer (abbreviated **Cnc**, Crab):

    clusters, open:

        M44 (Beehive Cluster, Manger, NGC 2632, Praesepe Cluster), *8h 40.1m, +19°59'*

        M67 (NGC 2682), *8h 50.4m, +11°49'*

    galaxies, Cancer Cluster, *8h 21.0m, +20°56'*

    stars:

        alpha (α) Cancri (Acubens), *8h 58.5m, +11°51'*

        delta (δ) Cancri (Asellus Australis), *8h 44.7m, +18°09'*

        gamma (γ) Cancri (Asellus Borealis), *8h 43.3m, +21°28'*

        zeta (ζ) Cancri (Tegmen), *8h 12.2m, +17°39'*

Cancer Cluster of galaxies, *8h 21.0m, +20°56'*

Canes Venatici (abbreviated **CVn**, Greyhounds, Hounds, Hunting Dogs):

cluster, globular, M3 (NGC 5272), *13h 42.2m, +28°23'*

galaxies:

Holmberg VIII, *13h 13.3m, +36°13'*

M51 (Lord Rosse's "Nebula," NGC 5194 and 5195, Question Mark Galaxy, Whirlpool Galaxy), *13h 29.9m, +47°12'* and *13h 30.0m, +47°16'*

M63 (NGC 5055, Sunflower Galaxy), *13h 15.8m, +42°02'*

M94 (NGC 4736), *12h 50.9m, +41°07'*

M106 (NGC 4258), *12h 19.0m, +47°18'*

stars:

alpha^2 (α^2) Canum Venaticorum (Cor Caroli), *12h 56.0m, +38°19'*

beta (β) Canum Venaticorum (Chara), *12h 33.7m, +41°21'*

Y Canum Venaticorum (La Superba), *12h 45.1m, +45°26'*

Canis Major (abbreviated **CMa**, Greater Dog):

clusters, open:

M41 (NGC 2287), *6h 47.0m, –20°44'*

omicron (o) Canis Majoris Cluster, *6h 54.2m, –24°38'*

stars:

alpha (α) Canis Majoris (Dog Star, Nile Star, Sirius), *6h 45.1m, –16°43'*

beta (β) Canis Majoris (Mirzam, Murzim), *6h 22.7m, –17°57'*

delta (δ) Canis Majoris (Wesen, Wezen), *7h 08.4m, –26°24'*

epsilon (ε) Canis Majoris (Adara, Adhara), *6h 58.6m, –28°58'*

eta (η) Canis Majoris (Aludra), *7h 24.1m, –29°18'*

gamma (γ) Canis Majoris (Muliphen), *7h 03.8m, –15°38'*

Sirius B (Pup), *6h 45.2m, –16°43'*

zeta (ζ) Canis Majoris (Furud, Phurud), *6h 20.3m, –30°04'*

Canis Minor (abbreviated **CMi**, Lesser Dog):

stars:

alpha (α) Canis Minoris (Little Dog Star, Procyon),
7h 39.3m, +5° 14'

beta (β) Canis Minoris (Gomeisa, Gomeiza, Gomelza),
7h 27.2m, +8° 17'

Huruhata's Object binary star, *7h 30.3m, +4° 32'*

Canopus (alpha (α) Carinae), *6h 24.0m, –52° 42'*

Cap. *See* **Capricornus**

Capella (alpha (α) Aurigae, Goat Star), *5h 16.7m, +46° 00'*

Caph (beta (β) Cassiopeiae), *0h 09.2m, +59° 09'*

Capricorn Dwarf System globular cluster in Capricornus,
21h 46.5m, –21° 14'

Capricornus (abbreviated **Cap**, Goat, Seagoat):

clusters, globular:

Capricorn Dwarf System, *21h 46.5m, –21° 14'*

M30 (NGC 7099), *21h 40.4m, –23° 11'*

stars:

alpha^2 (α^2) Capricorni (Algedi, Algiedi, Giedi), *20h 18.1m,
–12° 33'*

beta (β) Capricorni (Dabih), *20h 21.0m, –14° 47'*

delta (δ) Capricorni (Deneb Algedi), *21h 47.0m, –16° 08'*

gamma (γ) Capricorni (Nashira), *21h 40.1m, –16° 40'*

Caput Trianguli (alpha (α) Trianguli, Rasalmuthallath), *1h 53.1m,
+29° 35'*

Car. *See* **Carina**

Carafe Galaxy in Caelum, *4h 28.0m, –47° 54'*

Carafe Group of galaxies in Caelum (NGC 1595, 1598, and Carafe
Galaxy), *4h 28.4m, –47° 49'; 4h 28.6m, –47° 47'; and
4h 28.0m, –47° 54'*

Carina (abbreviated **Car**, Keel of Argo Navis):

cluster, open, theta (θ) Carinae (IC 2602, Southern Pleiades),
10h 43.2m, –64° 24'

Carina (cont.):

    galaxy, Carina Dwarf Galaxy, *6h 41.6m, –50°58'*

    nebulae, bright:

        Butterfly Nebula (IC 2220, Toby Jug Nebula), *7h 56.8m, –59°07'*

        eta (η) Carinae Nebula (NGC 3372), *10h 43.8m, –59°52'*

        Homunculus Nebula, *10h 43.8m, –59°52'*

    nebulae, dark, Keyhole Nebula, *10h 43.8m, –59°52'*

    stars:

        a Carinae, *9h 11.0m, –58°58'*

        alpha (α) Carinae (Canopus), *6h 24.0m, –52°42'*

        beta (β) Carinae (Miaplacidus), *9h 13.2m, –69°43'*

        epsilon (ε) Carinae (Avior), *8h 22.5m, –59°31'*

        epsilon & iota (ε & ι) Carinae and delta & kappa (δ & κ) Velorum (False Cross), *8h 22.5m, –59°31';* *9h 17.1m, –59°17'; 8h 44.7m, –54°42'; and 9h 22.1m, –55°01'*

        iota (ι) Carinae (Aspidiske), *9h 17.1m, –59°17'*

        l Carinae, *9h 45.2m, –62°30'*

        p Carinae, *10h 32.0m, –61°41'*

        q Carinae, *10h 17.1m, –61°20'*

        u Carinae, *10h 53.5m, –58°51'*

Carina Dwarf Galaxy, *6h 41.6m, –50°58'*

Cartwheel Galaxy in Sculptor, *0h 37.4m, –33°45'*

Cassiopeia (abbreviated **Cas**, Lady in the Chair):

    clusters, open:

        M52 (NGC 7654), *23h 24.2m, +61°35'*

        M103 (NGC 581), *1h 33.2m, +60°42'*

        Owl Cluster (NGC 457), *1h 19.1m, +58°20'*

    galaxies:

        Maffei I, *2h 36.3m, +59°39'*

        Maffei II, *2h 41.9m, +59°36'*

Cassiopeia (cont.):

    nebulae, bright:

        Bubble Nebula (NGC 7635), *23h 20.7m, +61° 12'*

        gamma (γ) Cassiopeiae Nebula (IC 59 and 63), *0h 56.7m, +61° 04' and 0h 59.5m, +60° 49'*

    stars:

        alpha (α) Cassiopeiae (Schedar, Schedir, Shedar), *0h 40.5m, +56° 32'*

        AO Cassiopeiae (Pearce's Star), *0h 17.7m, +51° 26'*

        B Cassiopeiae (Tycho's Star), *0h 25.3m, +64° 09'*

        beta (β) Cassiopeiae (Caph), *0h 09.2m, +59° 09'*

        Bradley 3077, *23h 13.3m, +57° 10'*

        delta (δ) Cassiopeiae (Ruchbah), *1h 25.8m, +60° 14'*

        gamma (γ) Cassiopeiae (Navi), *0h 56.7m, +60° 43'*

        Hoffmeister's Star, *23h 40.3m, +53° 58'*

        mu (μ) Cassiopeiae (Marfak, Marfik), *1h 08.3m, +54° 55'*

        Osawa's Star, *23h 32.8m, +57° 54'*

        theta (θ) Cassiopeiae (Marfak, Marfik), *1h 11.1m, +55° 09'*

Castor (alpha (α) Geminorum), *7h 34.6m, +31° 53'*

Castor and Pollux. *See* **Gemini**

Cave (bright) Nebula in Cepheus, *22h 56.8m, +62° 37'*

Cebalrai (beta (β) Ophiuchi, Cheleb), *17h 43.5m, +4° 34'*

Celaeno (16 Tauri), *3h 44.8m, +24° 17'*

Celaeno (bright) Nebula in Taurus, *3h 44.8m, +24° 17'*

Centaurus (abbreviated **Cen**, Centaur):

    cluster, globular, omega (ω) Centauri Cluster (NGC 5139), *13h 26.8m, –47° 29'*

    galaxies:

        Centaurus A (NGC 5128), *13h 25.5m, –43° 01'*

        Centaurus Chain, *12h 44.3m, –40° 43'*

        Centaurus Cluster, *12h 50.0m, –41° 18'*

        Fourcade-Figueroa Object, *13h 34.8m, –45° 33'*

Centaurus (cont.):

    galaxies (cont.):

        Hardcastle "Nebula," *13h 13.0m, –32°42'*

        Seashell Galaxy, *13h 47.4m, –30°25'*

    nebulae, bright:

        Boomerang Nebula, *12h 44.8m, –54°31'*

        lambda (λ) Centauri Nebula (IC 2944 and 2948),
            11h 36.6m, –63°02' and 11h 38.8m, –63°32'

        Running Chicken Nebula (IC 2944), *11h 36.6m, –63°02'*

    nebulae, planetary, Blue Nebula (NGC 3918), *11h 50.3m,*
        –57°11'

    stars:

        alpha^1 (α^1) Centauri (Rigil Kentaurus, Toliman),
            14h 39.6m, –60°50'

        beta (β) Centauri (Agena, Hadar, Wazn), *14h 03.8m,*
            –60°22'

        Bidelman's Helium Variable Star, *14h 23.0m, –39°31'*

        Fehrenbach's Star, *13h 26.5m, –47°16'*

        gamma (γ) Centauri (Muhlifain), *12h 41.5m, –48°58'*

        Krzeminski's Star, *11h 21.3m, –60°37'*

        Liller's Star, *11h 21.2m, –60°36'*

        Popper's Star, *14h 15.0m, –46°17'*

        Proxima Centauri (Inne's Star), *14h 29.7m, –62°41'*

        Przybylski's Star, *11h 37.6m, –46°43'*

        Schweizer-Middleditch Star, *15h 02.8m, –41°59'*

        theta (θ) Centauri (Menkent), *14h 06.7m, –36°22'*

Centaurus A galaxy (NGC 5128), *13h 25.5m, –43°01'*

Centaurus Chain of galaxies, *12h 44.3m, –40°43'*

Centaurus Cluster of galaxies, *12h 50.0m, –41°18'*

Cepheus (abbreviated **Cep**, King, Monarch, Warrior King):

    nebulae, bright:

        Cave Nebula, *22h 56.8m, +62°37'*

        Gyulbudaghian's Nebula, *20h 45.9m, +67°58'*

Cepheus (cont.):

    stars:

        alpha (α) Cephei (Alderamin), *21h 18.6m, +62°35'*

        beta (β) Cephei (Alfirk, Alphirk), *21h 28.7m, +70°34'*

        gamma (γ) Cephei (Alrai, Errai), *23h 39.3m, +77°38'*

        mu (μ) Cephei (Garnet Star, Herschel's Garnet Star), *21h 43.5m, +58°47'*

        xi (ξ) Cephei (Kurhah), *22h 03.8m, +64°38'*

Cetus (abbreviated **Cet**, Sea Monster, Whale):

    galaxies:

        Burbidge Chain, *0h 47.5m, –20°26'*

        Haufen A cluster, *1h 08.9m, –15°25'*

        M77 (NGC 1068), *2h 42.7m, –0°01'*

        Minkowski's Object, *1h 25.8m, –1°22'*

        Wolf-Lundmark-Melotte (WLM) System, *0h 02.0m, –15°28'*

    stars:

        alpha (α) Ceti (Menkab, Menkar), *3h 02.3m, +4°05'*

        beta (β) Ceti (Deneb Kaitos, Difda, Diphda, Second Frog), *0h 43.6m, –17°59'*

        Borrelly's Star, *0h 23.8m, –9°28'*

        eta (η) Ceti (Deneb, Deneb Algenubi), *1h 08.6m, –10°11'*

        iota (ι) Ceti (Deneb Kaitos, Schemali), *0h 19.4m, –8°49'*

        Luyten's Flare Star, *1h 38.8m, –17°58'*

        omicron (o) Ceti (Mira, Wonderful Star), *2h 19.3m, –2°59'*

        268 G. Ceti, *2h 36.1m, +6°53'*

        zeta (ζ) Ceti (Baten Kaitos), *1h 51.5m, –10°20'*

Cha (abbreviation for **Chamaeleon**, Chameleon)

Chained Lady. *See* **Andromeda**

Chamaeleon (abbreviated **Cha**, Chameleon)

Champion. *See* **Perseus**

Chanal's Variable Star in Orion, *5h 34.8m, –5°34'*

Chara (beta (β) Canum Venaticorum), *12h 33.7m, +41°21'*

Charioteer. *See* **Auriga**

Charles's Wain. *See* **Ursa Major**

Checkmark (bright) Nebula in Sagittarius (Horseshoe Nebula, M17, NGC 6618, Omega Nebula, Swan Nebula), *18h 20.8m, –16° 11'*

Cheleb (Cebalrai, beta (β) Ophiuchi), *17h 43.5m, +4° 34'*

Chemist's Furnace. *See* **Fornax**

Chertan (Chort, theta (θ) Leonis), *11h 14.2m, +15° 26'*

Chevremont's Star in Aquarius, *21h 33.5m, –0° 49'*

Chisel. *See* **Caelum**

Chort (Chertan, theta (θ) Leonis), *11h 14.2m, +15° 26'*

Christmas Tree (open) Cluster in Monoceros (in NGC 2264), *6h 41.1m, +9° 53'*

Chu's Object bright nebula in Perseus, *3h 56.8m, +51° 26'*

Circinus (abbreviated **Cir**, Compasses, Drawing Compasses, Pair of Compasses):

galaxy, Circinus Galaxy, *14h 13.2m, –65° 20'*

Circlet. *See* **Pisces**

Cirrus (bright) Nebula in Cygnus (Bridal Veil Nebula; NGC 6960, 6992, and 6995; Veil Nebula), *20h 45.7m, +30° 43';* *20h 56.4m, +31° 43'; and 20h 57.1m, +31° 13'*

Clock, Pendulum Clock (**Horologium**, abbreviated **Hor**)

Cloud (bright) Nebula in Scorpius (Antares Nebula), *16h 29.2m, –26° 27'*

Clown Face (planetary) Nebula in Gemini (Eskimo Nebula, NGC 2392), *7h 29.2m, +20° 55'*

CMa. *See* **Canis Major**

CMi. *See* **Canis Minor**

Cnc. *See* **Cancer**

Coalsack (dark) Nebula in Crux, *12h 53.0m, –63° 00'*

Coat Hanger open cluster in Vulpecula (Brocchi's Cluster), *19h 25.4m, +20° 11'*

Cocoon (bright) Nebula in Cygnus (IC 5146), *21h 53.4m, +47° 16'*

Coddington's "Nebula" galaxy in Ursa Major (IC 2574), *10h 28.4m, +68° 25'*

Cohen-Schwartz Star in Orion, *5h 36.4m, –6° 46'*

Col. *See* **Columba**

Colt. *See* **Equuleus**

Columba (abbreviated **Col**, Dove, Noah's Dove):

    stars:

        alpha (α) Columbae (Phact, Phaet), *5h 39.6m, –34° 04'*

        beta (β) Columbae (Wazn, Wezn), *5h 51.0m, –35° 46'*

Coma Berenices (abbreviated **Com**, Berenice's Hair):

    cluster, globular, M53 (NGC 5024), *13h 12.9m, +18° 10'*

    cluster, open, Coma Star Cluster (Melotte 111), *12h 25.0m,*
        +26° 00'

    galaxies:

        Blackeye Galaxy (M64, NGC 4826), *12h 56.7m, +21° 41'*

        Box galaxies (NGC 4169, 4173, 4174, and 4175), *12h 12.2m,*
            +29° 10'; 12h 12.3m, +29° 11'; 12h 12.4m, +29° 08';
            and *12h 12.5m, +29° 09'*

        Coma Cluster, *12h 59.8m, +27° 59'*

        M64 (Blackeye Galaxy, NGC 4826), *12h 56.7m, +21° 41'*

        M85 (NGC 4382), *12h 25.4m, +18° 11'*

        M88 (NGC 4501), *12h 32.0m, +14° 25'*

        M91—error in Messier Catalogue or possibly NGC 4548,
            12h 35.4m, +14° 30'

        M98 (NGC 4192), *12h 13.8m, +14° 54'*

        M99 (NGC 4254, Pinwheel Galaxy), *12h 18.8m, +14° 25'*

        M100 (NGC 4321), *12h 22.9m, +15° 49'*

        Mice (NGC 4676), *12h 46.1m, +30° 44'*

    star, Rosino-Zwicky variable star, *12h 32.4m, +14° 19'*

Coma Cluster of galaxies in Coma Berenices, *12h 59.8m, +27° 59'*

Coma Star (open) Cluster in Coma Berenices (Melotte 111),
    12h 25.0m, +26° 00'

Compasses:

    Compass of Argo Navis or Mariner's Compass (**Pyxis**,
        abbreviated **Pyx**)

    Drawing Compasses or Pair of Compasses. *See* **Circinus**

Cone (dark) Nebula in Monoceros (Conus Nebula, in NGC 2264), *6h 40.9m, +9°54'*

Copeland's Septet of galaxies in Leo (NGC 3745, 3746, 3748, 3750, 3751, 3753, and 3754), *11h 37.7m, +22°01'; 11h 37.7m, +22°00'; 11h 37.8m, +22°02'; 11h 37.9m, +21°58'; 11h 37.9m, +21°56'; 11h 37.9m, +21°59'; and 11h 37.9m, +21°59'*

Cor Caroli (alpha² (α²) Canum Venaticorum), *12h 56.0m, +38°19'*

Cor Hydrae (Alfard, alpha (α) Hydrae, Alphard, Red Bird), *9h 27.6m, –8°40'*

Cor Leonis (alpha (α) Leonis, Regulus), *10h 08.4m, +11°58'*

Cor Scorpii (alpha (α) Scorpii, Antares), *16h 29.4m, –26°26'*

Cor Serpentis (alpha (α) Serpentis, Unukalhai), *15h 44.3m, +6°26'*

Cor Tauri (alpha (α) Tauri, Aldebaran, Palilicium), *4h 35.9m, +16°31'*

Cork (planetary) Nebula in Perseus (Butterfly Nebula, Little Dumbbell Nebula, M76, NGC 650 and 651), *1h 42.3m, +51°34' and 1h 42.3m, +51°35'*

Corona Australis (abbreviated **CrA**, Southern Crown)

Corona Borealis (abbreviated **CrB**, Northern Crown):

galaxies, Corona Borealis Cluster, *15h 22.7m, +27°43'*

stars:

alpha (α) Coronae Borealis (Alfeta, Alphecca, Gemma), *15h 34.7m, +26°43'*

beta (β) Coronae Borealis (Nusakan), *15h 27.8m, +29°06'*

T Coronae Borealis (Blaze Star), *15h 59.5m, +25°55'*

Corona Borealis Cluster of galaxies, *15h 22.7m, +27°43'*

Corvus (abbreviated **Crv**, Crow, Raven):

galaxies, Antennae (NGC 4038 and 4039, Ring-Tail Galaxy, Snorter), *12h 01.9m, –18°52' and 12h 01.9m, –18°53'*

stars:

alpha (α) Corvi (Alchiba), *12h 08.4m, –24°44'*

delta (δ) Corvi (Algorab, Algores), *12h 29.9m, –16°31'*

gamma (γ) Corvi (Gienah), *12h 15.8m, –17°33'*

CrA (abbreviation for **Corona Australis,** Southern Crown)

Crab. *See* **Cancer**

Crab (bright) Nebula in Taurus (M1, NGC 1952), *5h 34.5m,
    +22° 01'*

Crane. *See* **Grus**

Crater (abbreviated **Crt**, Cup):

    stars:

        Abt's Star, *11h 17.0m, –7° 08'*

        alpha (α) Crateris (Alkes), *10h 59.8m, –18° 18'*

CrB. *See* **Corona Borealis**

Crescent (bright) Nebula in Cygnus (NGC 6888), *20h 12.0m,
    +38° 21'*

Crimson Star in Lepus (Hind's Crimson Star, R Leporis), *4h 59.6m,
    –14° 48'*

Crosses:

    False Cross (epsilon & iota (ε & ι) Carinae and delta & kappa
        (δ & κ) Velorum, *8h 22.5m, –59° 31'; 9h 17.1m,
        –59° 17'; 8h 44.7m, 54° 42'; and 9h 22.1m, –55° 01'*

    Northern Cross. *See* **Cygnus**

    Southern Cross. *See* **Crux**

Crow. *See* **Corvus**

Crowns:

    Northern Crown. *See* **Corona Borealis**

    Southern Crown (**Corona Australis**, abbreviated **CrA**)

Crt. *See* **Crater**

Crux (abbreviated **Cru**, Cross, Southern Cross):

    cluster, open, kappa (κ) Crucis Cluster (Jewel Box, NGC 4755),
        12h 53.6m, –60° 20'

    nebula, dark, Coalsack Nebula, *12h 53.0m, –63° 00'*

    stars:

        alpha¹ (α¹) Crucis (Acrux), *12h 26.6m, –63° 06'*

        beta (β) Crucis (Mimosa), *12h 47.7m, –59° 41'*

        gamma (γ) Crucis (Gacrux), *12h 31.2m, –57° 07'*

Crv. *See* **Corvus**

Cujam (omega (ω) Herculis), *16h 25.4m, +14°02'*

Cup. *See* **Crater**

Cursa (beta (β) Eridani, Dhalim, Kursa), *5h 07.8m, –5°05'*

CVn. *See* **Canes Venatici**

Cygnus (abbreviated **Cyg**, Northern Cross, Swan):

    clusters, open:

        M29 (NGC 6913), *20h 23.9m, +38°32'*

        M39 (NGC 7092), *21h 32.2m, +48°26'*

    nebulae, bright:

        America Nebula (NGC 7000, North America Nebula), *20h 58.8m, +44°20'*

        Bridal Veil Nebula (Cirrus Nebula; Cygnus Loop; NGC 6960, 6992, and 6995; Veil Nebula), *20h 45.7m, +30°43'; 20h 56.4m, +31°43'; and 20h 57.1m, +31°13'*

        Cocoon Nebula (IC 5146), *21h 53.4m, +47°16'*

        Crescent Nebula (NGC 6888), *20h 12.0m, +38°21'*

        Cygnus Superbubble, *20h 32.0m, +42°00'*

        Filamentary Nebula (Lacework Nebula, NGC 6960, part of Veil Nebula), *20h 45.7m, +30°43'*

        Footprint Nebula (Minkowski's Footprint), *19h 36.3m, +29°33'*

        gamma (γ) Cygni Nebula (Cygnus Nebula, IC 1318), *20h 22.2m, +40°15'*

        Network Nebula (NGC 6992 and 6995, part of Veil Nebula), *20h 56.4m, +31°43' and 20h 57.1m, +31°13'*

        Pelican Nebula (IC 5067 and 5070), *20h 47.8m, +44°22' and 20h 50.8m, +44°21'*

        Pickering's Triangular Wisp (part of Veil Nebula), *20h 48.5m, +31°09'*

    nebulae, dark:

        Fish on the Platter Nebula, *19h 59.0m, +35°00'*

        Gulf of Mexico, *20h 58.8m, +44°20'*

        Northern Coalsack Nebula, *20h 40.0m, +42°00'*

Cygnus (cont.):

nebulae, planetary:

Blinking Planetary Nebula (NGC 6826), *19h 44.8m, +50°31'*

Cygnus Egg (Egg Nebula), *21h 02.3m, +36°42'*

stars:

alpha (α) Cygni (Arided, Deneb), *20h 41.4m, +45°17'*

alpha (α) Cygni, alpha (α) Aquilae, and alpha (α) Lyrae (Summer Triangle), *20h 41.4m, +45°17'; 19h 50.8m, +8°52'; and 18h 36.9m, +38°47'*

beta (β) Cygni (Albireo), *19h 30.7m, +27°58'*

Campbell's Star, *19h 34.8m, +30°31'*

epsilon (ε) Cygni (Gienah), *20h 46.2m, +33°58'*

Flying Star (Piazzi's Flying Star, 61 Cygni), *21h 06.9m, +38°45'*

gamma (γ) Cygni (Sadr), *20h 22.2m, +40°15'*

Honda's Variable Star, *21h 43.9m, +31°28'*

P Cygni (34 Cygni), *20h 17.8m, +38°02'*

pi¹ (π¹) Cygni (Azelfafage), *21h 42.1m, +51°11'*

Cygnus Egg planetary nebula (Egg Nebula), *21h 02.3m, +36°42'*

Cygnus Loop bright nebula (Bridal Veil Nebula; Cirrus Nebula; NGC 6960, 6992, and 6995; Veil Nebula), *20h 45.7m, +30°43'; 20h 56.4m, +31°43'; and 20h 57.1m, +31°13'*

Cygnus (bright) Nebula (gamma (γ) Cygni Nebula, IC 1318), *20h 22.2m, +40°15'*

Cygnus Superbubble bright nebula, *20h 32.0m, +42°00'*

Cynosura (alpha (α) Ursae Minoris, North Star, Polaris, Pole Star), *2h 31.8m, +89°16'* d Bootis (12 Bootis), *14h 10.4m, +25° 06'*

d Serpentis (59 Serpentis), *18h 27.2m, +0° 12'*

Dabih (beta (β) Capricorni), *20h 21.0m, –14° 47'*

Dafira (beta (β) Leonis, Deneb, Denebola, Serpha), *11h 49.1m, +14° 34'*

Dark Bay (dark) Nebula in Orion (Horsehead Nebula), *5h 41.0m, –2° 24'*

Dark S (dark) Nebula in Ophiuchus (Barnard's S Nebula, S Nebula, Snake Nebula), *17h 23.5m, –23° 38'*

Darklane Galaxy in Virgo (M104, NGC 4594, Sombrero Galaxy), *12h 40.0m, –11° 37'*

Das Rheingold galaxy in Volans (Graham's Object, Niebelungen Ring), *6h 41.4m, –74° 19'*

Del. *See* **Delphinus**

Delle Caustiche star cloud in Sagittarius (M24, Small Sagittarius Cloud), *18h 16.9m, –18° 29'*

Delphinus (abbreviated **Del**, Dolphin, Job's Coffin):

 stars:

  alpha (α) Delphini (Sualocin, Svalocin), *20h 39.6m, +15° 55'*

  beta (β) Delphini (Rotanev), *20h 37.5m, +14° 36'*

  epsilon (ε) Delphini (Deneb), *20h 33.2m, +11° 18'*

delta (δ) Aquarii (Scheat, Skat), *22h 54.6m, –15° 49'*

delta (δ) Cancri (Asellus Australis), *8h 44.7m, +18° 09'*

delta (δ) Canis Majoris (Wesen, Wezen), *7h 08.4m, –26° 24'*

delta (δ) Capricorni (Deneb Algedi), *21h 47.0m, –16° 08'*

delta (δ) Cassiopeiae (Ruchbah), *1h 25.8m, +60° 14'*

delta (δ) Corvi (Algorab, Algores), *12h 29.9m, –16° 31'*

delta (δ) Draconis (Altais, Nodus Secundus), *19h 12.6m, +67° 40'* delta (δ) Geminorum (Wasat), *7h 20.1m, +21° 59'*

delta (δ) Leonis (Dhur, Duhr, Zosma, Zubra), *11h 14.1m, +20° 31'*

delta (δ) Lyrae (open) Cluster, *18h 53.5m, +36° 55'*

delta (δ) Ophiuchi (Yed Prior), *16h 14.3m, –3° 42'*

delta (δ) Orionis (Mintaka), *5h 32.0m, –0° 18'*

delta, epsilon, and zeta (δ, ε, and ζ) Orionis (Belt of Orion),
      5h 32.0m, –0° 18'; 5h 36.2m, –1° 12'; and 5h 40.8m, –1° 57'

delta (δ) Sagittarii (Media, Kaus Meridionalis), *18h 21.0m, –29° 50'*

delta (δ) Scorpii (Dschubba), *16h 00.3m, –22° 37'*

delta¹ (δ¹) Tauri (Hyadum II), *4h 22.9m, +17° 33'*

delta (δ) Ursae Majoris (Megrez), *12h 15.4m, +57° 02'*

delta (δ) Ursae Minoris (Yildun), *17h 32.2m, +86° 35'*

delta & kappa (δ & κ) Velorum and epsilon & iota (ε & ι) Carinae
      (False Cross), *8h 44.7m, –54° 42'; 9h 22.1m, –55° 01';
      8h 22.5m, –59° 31'; and 9h 17.1m, –59° 17'*

Demon Star (Algol, beta (β) Persei, Medusa Star), *3h 08.2m,
      +40° 57'*

Deneb:

   alpha (α) Cygni (Arided), *20h 41.4m, +45° 17'*

   beta (β) Leonis (Dafira, Denebola, Serpha), *11h 49.1m, +14° 34'*

   epsilon (ε) Aquilae, *18h 59.6m, +15° 04'*

   epsilon (ε) Delphini, *20h 33.2m, +11° 18'*

   eta (η) Ceti (Deneb Algenubi), *1h 08.6m, –10° 11'*

   zeta (ζ) Aquilae, *19h 05.4m, +13° 52'*

Deneb Algedi (delta (δ) Capricorni), *21h 47.0m, –16° 08'*

Deneb Algenubi (Deneb, eta (η) Ceti), *1h 08.6m, –10° 11'*

Deneb Kaitos:

   beta (β) Ceti (Difda, Diphda, Second Frog), *0h 43.6m, –17° 59'*

   iota (ι) Ceti (Schemali), *0h 19.4m, –8° 49'*

Denebola (beta (β) Leonis, Dafira, Deneb, Serpha), *11h 49.1m,
      +14° 34'*

Dhalim (beta (β) Eridani, Cursa, Kursa), *5h 07.8m, –5° 05'*

Dhur (delta (δ) Leonis, Duhr, Zosma, Zubra), *11h 14.1m, +20° 31'*

Diabolo (planetary) Nebula in Vulpecula (Double-Headed Shot,
    Dumbbell Nebula, M27, NGC 6853), *19h 59.6m, +22° 43'*

Difda (beta (β) Ceti, Deneb Kaitos, Diphda, Second Frog),
    0h 43.6m, –17° 59'

Dippers:

   Big Dipper. *See* **Ursa Major**

   Little Dipper. *See* **Ursa Minor**

   Milk Dipper. *See* **Sagittarius**

Dnoces (iota (ι) Ursae Majoris, Talita, Talitha), *8h 59.2m, +48° 02'*

Dog Star (alpha (α) Canis Majoris, Nile Star, Sirius), *6h 45.1m,
    –16° 43'*

Dogs:

   Greater Dog. *See* **Canis Major**

   Greyhounds, Hounds, or Hunting Dogs. *See* **Canes Venatici**

   Lesser Dog. *See* **Canis Minor**

Dolphin. *See* **Delphinus**

Dorado (abbreviated **Dor**, Goldfish, Swordfish):

   galaxy, Large Magellanic Cloud (Great Magellanic Cloud,
       LMC, Nubecula Major), centered at *5h 23.6m, –69° 45'*

   nebulae, bright:

       Seagull Nebula (NGC 2032, 2033, 2034, and 2035),
           *5h 35.3m, –67° 34'; 5h 34.5m, –69° 44m; 5h 35.7m,
           –66° 56'; and 5h 35.3m, –67° 34'*

       30 Doradus Nebula (Great Looped Nebula, Loop Nebula,
           NGC 2070, Tarantula Nebula), *5h 38.7m, –69° 06'*

   stars:

       Walborn's Star, *4h 55.3m, –67° 11'*

       Warren and Penfold's (WP) Star, *5h 39.0m, –64° 05'*

Double (open) Cluster in Perseus (NGC 869 and 884, Sword-Hand
    of Perseus), *2h 19.0m, +57° 09'* and *2h 22.4m, +57° 07'*

Double-Double star in Lyra (epsilon[1] and epsilon[2] (ε[1] and ε[2]) Lyrae),
    18h 44.3m, +39° 40' and *18h 44.4m, +39° 37'*

double-double star in Serpens (Tweedledee and Tweedledum),
    18h 45.5m, +5° 30'

Double-Headed Shot planetary nebula in Vulpecula (Diabolo
Nebula, Dumbbell Nebula, M27, NGC 6853), *19h 59.6m,
+22° 43'*

Dove. *See* **Columba**

Draco (abbreviated **Dra**, Dragon):

galaxy, Draco Dwarf Galaxy, *17h 20.2m, +57° 55'*

stars:

alpha (α) Draconis (Thuban), *14h 04.4m, +64° 23'*

beta (β) Draconis (Alwaid, Rastaban), *17h 30.4m, +52° 18'*

delta (δ) Draconis (Altais, Nodus Secundus), *19h 12.6m,
+67° 40'*

gamma (γ) Draconis (Eltanin, Etamin, Ettanin, Rasaben,
Rastaban), *17h 56.6m, +51° 29'*

iota (ι) Draconis (Edasich), *15h 24.9m, +58° 58'*

lambda (λ) Draconis (Giansar or Giauzar), *11h 31.4m,
+69° 20'*

mu (μ) Draconis (Alrakis, Arrakis), *17h 05.3m, +54° 28'*

psi (ψ) Draconis (Dsiban), *17h 41.9m, +72° 09'*

Roberts-Altizer variable star, *10h 15.6m, +73° 26'*

xi (ξ) Draconis (Grumium), *17h 53.5m, +56° 52'*

Draco Dwarf Galaxy, *17h 20.2m, +57° 55'*

Dragon. *See* **Draco**

Dragon dark nebula in Sagittarius, *18h 04.8m, –24° 30'*

Drawing Compasses. *See* **Circinus**

Dschubba (delta (δ) Scorpii), *16h 00.3m, –22° 37'*

Dsiban (psi (ψ) Draconis), *17h 41.9m, +72° 09'*

Dubhe (alpha (α) Ursae Majoris), *11h 03.7m, +61° 45'*

Duhr (delta (δ) Leonis, Dhur, Zosma, Zubra), *11h 14.1m, +20° 31'*

Dumbbell (planetary) Nebula in Vulpecula (Diabolo Nebula,
Double-Headed Shot, M27, NGC 6853), *19h 59.6m, +22° 43'*

Eagle. *See* **Aquila**

Eagle (bright) Nebula:

    IC 2177 in Monoceros, *7 h 05.1m, –10° 42'*

    M16 in Serpens (NGC 6611, Star Queen Nebula), *18h 18.8m, –13° 47'*

Easel. *See* **Pictor**

Edasich (iota (ι) Draconis), *15h 24.9m, +58° 58'*

Egg (planetary) Nebula in Cygnus (Cygnus Egg), *21h 02.3m, +36° 42'*

Eggen's Nearby Star in Sculptor, *1h 32.3m, –30° 41'*

Eight-burst (planetary) Nebula in Vela (NGC 3132), *10h 07.0m, –40° 26'*

82 G. Eridani, *3h 19.9m, –43° 04'*

El Alcola (Alula Australis, xi (ξ) Ursae Majoris), *11h 18.2m, +31° 32'*

Electra (17 Tauri), *3h 44.9m, +24° 07'*

Electra (bright) Nebula in Taurus, *3h 44.9m, +24° 07'*

Elgebar (Algebar, beta (β) Orionis, Rigel), *5h 14.5m, –8° 12'*

Elnath (Alnath, beta (β) Tauri, Nath), *5h 26.3m, +28° 36'*

Eltanin (Etamin, Ettanin, gamma (γ) Draconis, Rasaben, Rastaban), *17h 56.6m, +51° 29'*

Engraving Tool. *See* **Caelum**

Enif (epsilon (ε) Pegasi, Fumalfaras), *21h 44.2m, +9° 52'*

epsilon (ε) Aquilae (Deneb), *18h 59.6m, +15° 04'*

epsilon (ε) Aquarii (Albali), *20h 47.7m, –9° 30'*

epsilon, eta, and zeta (ε, η, and ζ) Aurigae (Kids), *5h 02.0m, +43° 49'; 5h 06.5m, +41° 14'; and 5h 02.5m, +41° 05'*

epsilon (ε) Bootis (Izar, Mizar, Pulcherrima), *14h 45.0m, +27° 04'* epsilon (ε) Canis Majoris (Adara, Adhara), *6h 58.6m, –28° 58'*

epsilon (ε) Carinae (Avior), *8h 22.5m, –59° 31'*

epsilon & iota (ε & ι) Carinae and delta & kappa (δ & κ) Velorum (False Cross), *8h 22.5m. –59° 31'; 9h 17.1m, –59° 17'; 8h 44.7m, –54° 42'; and 9h 22.1m, –55° 01'*

epsilon (ε) Cygni (Gienah), *20h 46.2m, +33° 58'*

epsilon (ε) Delphini (Deneb), *20h 33.2m, +11° 18'*

epsilon (ε) Geminorum (Meboula, Mebsuta, Mebusta), *6h 43.9m, +25° 08'*

epsilon (ε) Leonis (Algenubi), *9h 45.8m, +23° 46'*

epsilon[1] and epsilon[2] (ε^1 and ε^2) Lyrae (Double-Double), *18h 44.3m, +39° 40'* and *18h 44.4m, +39° 37'*

epsilon (ε) Ophiuchi (Yed Posterior), *16h 18.3m, –4° 42'*

epsilon (ε) Orionis (Alnilam, Anilam), *5h 36.2m, –1° 12'*

epsilon (ε) Pegasi (Enif, Fumalfaras), *21h 44.2m, +9° 52'*

epsilon (ε) Sagittarii (Kaus Australis), *18h 24.2m, –34° 23'*

epsilon (ε) Tauri (Ain), *4h 28.6m, +19° 11'*

epsilon (ε) Ursae Majoris (Alioth), *12h 54.0m, +55° 58'*

epsilon (ε) Virginis (Vindemiatrix), *13h 02.2m, +10° 58'*

Equuleus (abbreviated **Equ**, Colt, Foal, Little Horse):
> star, alpha (α) Equulei (Kitalpha), *21h 15.8m, +5° 15'*

Eridanus (abbreviated **Eri**, River):
> cluster, globular, Eridanus Cluster, *4h 24.8m, –21° 11'*
> galaxy, Holmberg VI, *3h 24.8m, –21° 20'*
> nebula, bright, Witch Head Nebula (IC 2118), *5h 06.9m, –7° 13'*
> stars:
>> alpha (α) Eridani (Achernar), *1h 37.7m, –57° 14'*
>> beta (β) Eridani (Cursa, Dhalim, Kursa), *5h 07.8m, –5° 05'*
>> 82 G. Eridani, *3h 19.9m, –43° 04'*
>> eta (η) Eridani (Azha), *2h 56.4m, –8° 54'*
>> gamma (γ) Eridani (Zaurak), *3h 58.0m, –13° 31'*
>> omicron[1] (o^1) Eridani (Beid), *4h 11.9m, –6° 50'*
>> omicron[2] (o^2) Eridani (Keid), *4h 15.3m, –7° 39'*

Eridanus (cont.):

> stars (cont.):
>
>> p Eridani, *1h 39.8m, –56° 12'*
>>
>> theta^1 (θ^1) Eridani (Acamar), *2h 58.3m, –40° 18'*

Eridanus (globular) Cluster, *4h 24.8m, –21° 11'*

Errai (Alrai, gamma (γ) Cephei), *23h 39.3m, +77° 38'*

Eskimo (planetary) Nebula in Gemini (Clown Face Nebula, NGC 2392), *7h 29.2m, +20° 55'*

eta (η) Bootis (Muphrid), *13h 54.7m, +18° 24'*

eta (η) Canis Majoris (Aludra), *7h 24.1m, –29° 18'*

eta (η) Carinae (bright) Nebula (NGC 3372), *10h 43.8m, –59° 52'*

eta (η) Ceti (Deneb, Deneb Algenubi), *1h 08.6m, –10° 11'*

eta (η) Eridani (Azha), *2h 56.4m, –8° 54'*

eta (η) Geminorum (Propus), *6h 14.9m, +22° 30'*

eta (η) Ophiuchi (Sabik), *17h 10.4m, –15° 44'*

eta (η) Pegasi (Matar, Sadalmatar), *22h 43.0m, +30° 13'*

eta (η) Tauri (Alcyone), *3h 47.5m, +24° 06'*

eta (η) Ursae Majoris (Alcaid, Alkaid, Benetnasch), *13h 47.5m, +49° 19'*

eta (η) Virginis (Zaniah), *12h 19.9m, –0° 40'*

Etamin (Eltanin, Ettamin, gamma (γ) Draconis, Rasaben, Rastaban), *17h 56.6m, +51° 29'*

Exclamation Mark Galaxy in Phoenix, *0h 39.3m, –43° 06'*

Eyes galaxies in Virgo (NGC 4435 and 4438), *12h 27.7m, +13° 05'* and *12h 27.8m, +13° 01'*

Ff

f Aquilae (26 Aquilae), *19h 20.5m, –5° 25'*

False Cross (epsilon & iota (ε & ι) Carinae and delta & kappa
(δ & κ) Velorum, *8h 22.5m, –59° 31'; 9h 17.1m, –59° 17';
8h 44.7m, –54° 42'; and 9h 22.1m, –55° 01'*

Fath 703 galaxy in Libra, *15h 13.8m, –15° 28'*

Fehrenbach's Star in Centaurus, *13h 26.5m, –47° 16'*

Field of the Nebulae galaxies in Virgo (Virgo Cluster), *12h 30.0m,
+12° 23'*

Filamentary (bright) Nebula in Cygnus (Lacework Nebula, NGC
6960, part of Veil Nebula), *20h 45.7m, +30° 43'*

First Frog (alpha (α) Piscis Austrini, Fomalhaut), *22h 57.6m,
–29° 37'*

Fish:

> Dolphin. *See* **Delphinus**
>
> Flying Fish. *See* **Volans**
>
> Goldfish. *See* **Dorado**
>
> Southern Fish. *See* **Piscis Austrinus**
>
> Swordfish. *See* **Dorado**

Fish on the Platter (dark) Nebula in Cygnus, *19h 59.0m, +35° 00'*

Fishes. *See* **Pisces**

Fish's Mouth dark nebula in Orion, *5h 35.4m, –5° 23'*

Flaming Star (bright) Nebula in Auriga (IC 405, Schaeberle's
Flaming Star), *5h 16.2m, +34° 16'*

Fly (**Musca**, abbreviated **Mus**)

Flying Fish. *See* **Volans**

Flying Star:

> Groombridge 1830 in Ursa Major (Runaway Star), *11h 53.0m,
> +37° 43'*
>
> Piazzi's Flying Star in Cygnus (61 Cygni), *21h 06.9m, +38° 45'*

Foal. *See* **Equuleus**

Fomalhaut (alpha (α) Piscis Austrini, First Frog), *22h 57.6m, –29°37'*

Footprint (bright) Nebula in Cygnus (Minkowski's Footprint), *19h 36.3m, +29°33'*

Fornax (abbreviated **For**, Chemist's Furnace, Furnace):

galaxies:

Fornax A Galaxy (NGC 1316), *3h 22.7m, –37°12'*

Fornax Dwarf Galaxy, *2h 39.9m, –34°32'*

Fornax I Cluster, *3h 32.0m, –35°20'*

Fornax II Cluster, *3h 28.0m, –20°45'*

47 Tucanae globular cluster (NGC 104), *0h 24.1m, –72°05'*

Fourcade-Figueroa Object galaxy in Centaurus, *13h 34.8m, –45°33'*

Fox. *See* **Vulpecula**

Fumalfaras (Enif, epsilon (ε) Pegasi), *21h 44.2m, +9°52'*

Furnace. *See* **Fornax**

Furud (Phurud, zeta (ζ) Canis Majoris), *6h 20.3m, –30°04'*

Gg

g Herculis (30 Herculis), *16h 28.6m, +41°53'*

G Scorpii, *17h 49.9m, –37°03'*

g Ursae Majoris (Alcor, 80 Ursae Majoris, Saidak), *13h 25.2m, +54°59'*

Gacrux (gamma (γ) Crucis), *12h 31.2m, –57°07'*

gamma[1] (γ^1) Andromedae (Almach, Almak), *2h 03.9m, +42°20'*

gamma (γ) Aquarii (Sadachbia), *22h 21.7m, –1°23'*

gamma (γ) Aquilae (Tarazed), *19h 46.3m, +10°37'*

gamma[2] (γ^2) Arietis (Mesartim), *1h 53.5m, +19°18'*

gamma (γ) Bootis (Seginus), *14h 32.1m, +38°19'*

gamma (γ) Cancri (Asellus Borealis), *8h 43.3m, +21°28'*

gamma (γ) Canis Majoris (Muliphen), *7h 03.8m, –15°38'*

gamma (γ) Capricorni (Nashira), *21h 40.1m, –16°40'*

gamma (γ) Cassiopeiae (Navi), *0h 56.7m, +60°43'*

gamma (γ) Cassiopeiae (bright) Nebula (IC 59 and 63), *0h 56.7m, +61°04'* and *0h 59.5m, +60°49'*

gamma (γ) Centauri (Muhlifain), *12h 41.5m, –48°58'*

gamma (γ) Cephei (Alrai, Errai), *23h 39.3m, +77°38'*

gamma (γ) Corvi (Gienah), *12h 15.8m, –17°33'*

gamma (γ) Crucis (Gacrux), *12h 31.2m, –57°07'*

gamma (γ) Cygni (Sadr), *20h 22.2m, +40°15'*

gamma (γ) Cygni (bright) Nebula (Cygnus Nebula, IC 1318), *20h 22.2m, +40°15'*

gamma (γ) Draconis (Eltanin, Etamin, Ettanin, Rasaben, Rastaban), *17h 56.6m, +51°29'*

gamma (γ) Eridani (Zaurak), *3h 58.0m, –13°31'*

gamma (γ) Geminorum (Alhena, Almeisam), *6h 37.7m, +16°24'*

gamma[1] (γ^1) Leonis (Algeiba, Algieba), *10h 20.0m, +19°50'* gamma (γ) Lyrae (Sulafat, Sulaphat), *18h 58.9m, +32° 41'*

gamma (γ) Orionis (Amazon Star, Bellatrix), *5h 25.1m, +6° 21'*

gamma (γ) Pegasi (Algenib), *0h 13.2m, +15° 11'*

gamma (γ) Sagittarii (Alnasl), *18h 05.8m, –30° 25'*

gamma (γ) Tauri (Hyadum I), *4h 19.8m, +15° 38'*

gamma (γ) Ursae Majoris (Phad, Phaed, Phecda), *11h 53.8m, +53° 42'*

gamma (γ) Ursae Minoris (Pherkad), *15h 20.7m, +71° 50'*

gamma^2 (γ^2) Velorum (Regor, Spectral Gem of the Southern Skies, Suhail), *8h 09.5m, –47° 20'*

gamma (γ) Virginis (Antevorta, Porrima), *12h 41.7m, –1° 27'*

Garnet Star in Cepheus (Herschel's Garnet Star, mu (μ) Cephei), *21h 43.5m, +58° 47'*

Gem. *See* **Gemini**

Gem of the Milky Way (Scutum Star Cloud), *18h 40.0m, –7° 00'*

Gemini (abbreviated **Gem**, Castor and Pollux, Twins):

    cluster, open, M35 (NGC 2168), *6h 08.9m, +24° 20'*

    galaxies:

        Gemini Cluster, *7h 07.6m, +35° 03'*

        Snickers, *6h 28.0m, +15° 00'*

    nebulae, planetary:

        Clown Face Nebula (Eskimo Nebula, NGC 2392), *7h 29.2m, +20° 55'*

        Medusa Nebula, *7h 29.0m, +13° 15'*

    stars:

        alpha (α) Geminorum (Castor), *7h 34.6m, +31° 53'*

        beta (β) Geminorum (Pollux), *7h 45.3m, +28° 02'*

        delta (δ) Geminorum (Wasat), *7h 20.1m, +21° 59'*

        epsilon (ε) Geminorum (Meboula, Mebsuta, Mebusta), *6h 43.9m, +25° 08'*

        eta (η) Geminorum (Propus), *6h 14.9m, +22° 30'*

        gamma (γ) Geminorum (Alhena, Almeisam), *6h 37.7m, +16° 24'*

Gemini (cont.):

    stars (cont.):

        mu (μ) Geminorum (Tejat), *6h 23.0m, +22°31'*

        zeta (ζ) Geminorum (Mekbuda), *7h 04.1m, +20°34'*

Gemini Cluster of galaxies, *7h 07.6m, +35°03'*

Gemma (Alfeta, alpha (α) Coronae Borealis, Alphecca), *15h 34.7m, +26°43'*

Ghost of Jupiter planetary nebula in Hydra (Jupiter Nebula, NGC 3242), *10h 24.8m, –18°38'*

Giansar or **Giauzar** (lambda (λ) Draconis), *11h 31.4m, +69°20'*

Giant. *See* **Orion**

Gibson Reaves 8 (GR 8) galaxy in Virgo, *12h 58.7m, +14°13'*

Giedi (Algedi, Algiedi, alpha^2 (α^2) Capricorni), *20h 18.1m, –12°33'*

Gienah:

    epsilon (ε) Cygni, *20h 46.2m, +33°58'*

    gamma (γ) Corvi, *12h 15.8m, –17°33'*

Giraffe. *See* **Camelopardalis**

Goat. *See* **Capricornus**

Goat Star (alpha (α) Aurigae, Capella), *5h 16.7m, + 46°00'*

Goldfish. *See* **Dorado**

Gomeisa (beta (β) Canis Minoris, Gomeiza, Gomelza), *7h 27.2m, +8°17'*

Graffias (Acrab, Akrab, beta^1 (β^1) Scorpii), *16h 05.4m, –19°48'*

Graham's Object galaxy in Volans (Das Rheingold, Niebelungen Ring), *6h 41.4m, –74°19'*

Gravingtool. *See* **Caelum**

Great (globular) Cluster in Hercules (Hercules Cluster, M13, NGC 6205), *16h 41.7m, +36°28'*

Great Galaxy in Andromeda (Andromeda Galaxy, M31, NGC 224), *0h 42.7m, +41°16'*

Great Looped (bright) Nebula in Dorado (Loop Nebula, NGC 2070, Tarantula Nebula, 30 Doradus Nebula), *5h 38.7m, –69°06'*

Great Magellanic Cloud galaxy in Dorado and Mensa (Large
Magellanic Cloud, LMC, Nubecula Major), centered in
Dorado at *5h 23.6m, –69°45'*

Great (bright) Nebula in Orion (M42, NGC 1976, Orion Nebula),
5h 35.4m, –5°27'

Great Square of Pegasus (alpha, beta, & gamma (α, β, & γ) Pegasi
and alpha (α) Andromedae), *23h 04.8m, +15°12';
23h 03.8m, +28°05'; & 0h 13.2m, +15°11'; and
0h 08.4m, +29°05'*

Greater Bear. *See* **Ursa Major**

Greater Dog. *See* **Canis Major**

Greyhounds. *See* **Canes Venatici**

Groombridge 1618 star in Ursa Major, *10h 11.4m, +49°27'*

Groombridge 1830 star in Ursa Major (Flying Star, Runaway Star),
11h 53.0m, +37°43'

Gru. *See* **Grus**

Grumium (xi (ξ) Draconis), *17h 53.5m, +56°52'*

Grus (abbreviated **Gru**, Crane):

galaxies, Grus Quartet (NGC 7552, 7582, 7590, and 7599),
*23h 16.2m, –42°35'; 23h 18.4m, –42°22'; 23h 18.9m,
–42°14'; and 23h 19.3m, –42°15'*

star, alpha (α) Gruis (Alnair), *22h 08.2m, –46°58'*

Grus Quartet of galaxies (NGC 7552, 7582, 7590, and 7599),
*23h 16.2m, –42°35'; 23h 18.4m, –42°22'; 23h 18.9m,
–42°14'; and 23h 19.3m, –42°15'*

Gulf of Mexico dark nebula in Cygnus, *20h 58.8m, +44°20'*

Gum (bright) Nebula in Vela, *8h 30.0m, –45°00'*

Gyulbudaghian's (bright) Nebula in Cepheus, *20h 45.9m, +67°58'*

Hh

Hadar (Agena, beta (β) Centauri, Wazn), *14h 03.8m, –60° 22'*

Hamal (alpha (α) Arietis), *2h 07.2m, +23° 28'*

Hammerhead (bright) Nebula in Orion, *5h 33.7m, –5° 40'*

Hardcastle "Nebula" galaxy in Centaurus, *13h 13.0m, –32° 42'*

Hare. *See* **Lepus**

Harp. *See* **Lyra**

Harp Star (alpha (α) Lyrae, Vega, Wega), *18h 36.9m, +38° 47'*

Haufen A cluster of galaxies in Cetus, *1h 08.9m, –15° 25'*

Helical (planetary) Nebula in Aquarius (Helix Nebula, NGC 7293), *22h 29.6m, –20° 48'*

Helix Galaxy in Ursa Major (NGC 2685), *8h 55.6m, +58° 44'*

Helix (planetary) Nebula in Aquarius (Helical Nebula, NGC 7293), *22h 29.6m, –20° 48'*

Hercules (abbreviated **Her**, Keystone, Kneeler):

clusters, globular:

M13 (Great Cluster in Hercules, Hercules Cluster, NGC 6205), *16h 41.7m, +36° 28'*

M92 (NGC 6341), *17h 17.1m, +43° 08'*

galaxies:

Hercules A, *16h 51.1m, +5° 00'*

Hercules Cluster, *16h 05.2m, +17° 45'*

Zwicky's Triplet, *16h 49.5m, +45° 30'*

stars:

alpha^1 (α^1) Herculis (Rasalgethi), *17h 14.6m, +14° 23'*

beta (β) Herculis (Kornephoros), *16h 30.2m, +21° 29'*

g Herculis (30 Herculis), *16h 28.6m, +41° 53'*

kappa (κ) Herculis (Marfak, Marfik), *16h 08.1m, +17° 03'*

lambda (λ) Herculis (Masym), *17h 30.7m, +26° 07'*

omega (ω) Herculis (Cujam), *16h 25.4m, +14° 02'*

Hercules (cont.):

  stars (cont.):

      Sanduleak-Pesch star, *17h 05.5m, +48°03'*

      u Herculis (68 Herculis), *17h 17.3m, +33°06'*

Hercules A Galaxy, *16h 51.1m, +5°00'*

Hercules (globular) Cluster (Great Cluster in Hercules, M13, NGC 6205), *16h 41.7m, +36°28'*

Hercules Cluster of galaxies, *16h 05.2m, +17°45'*

Herdsman. *See* **Bootes**

Herschel's Garnet Star in Cepheus (Garnet Star, mu (μ) Cephei), *21h 43.5m, +58°47'*

Hind's Crimson Star in Lepus (Crimson Star, R Leporis), *4h 59.6m, –14°48'*

Hind's New Star in Ophiuchus, *16h 59.5m, –12°54'*

Hind's Variable (bright) Nebula in Taurus (NGC 1555), *4h 22.9m, +19°32'*

Hoffmeister's Cloud dark nebula in Microscopium, *20h 47.0m, –42°00'*

Hoffmeister's Star in Cassiopeia, *23h 40.3m, +53°58'*

Holmberg I galaxy in Ursa Major, *9h 40.5m, +71°11'*

Holmberg II galaxy in Ursa Major, *8h 18.9m, +70°43'*

Holmberg III galaxy in Camelopardalis, *9h 14.8m, +74°14'*

Holmberg IV galaxy in Ursa Major, *13h 54.8m, +53°54'*

Holmberg V galaxy in Ursa Major, *13h 40.7m, +54°20'*

Holmberg VI galaxy in Eridanus, *3h 24.8m, –21°20'*

Holmberg VII galaxy in Virgo, *12h 34.7m, +6°18'*

Holmberg VIII galaxy in Canes Venatici, *13h 13.3m, +36°13'*

Holmberg IX galaxy in Ursa Major, *9h 57.6m, +69°03'*

Homam (zeta (ζ) Pegasi), *22h 41.5m, +10°50'*

Homunculus (bright) Nebula in Carina, *10h 43.8m, –59°52'*

Honda's Variable Star in Cygnus, *21h 43.9m, +31°28'*

Horologium (abbreviated **Hor**, Clock, Pendulum Clock)

Horsehead (dark) Nebula in Orion (Dark Bay Nebula), *5h 41.0m, –2°24'*

Horses:

Colt, Foal, or Little Horse. *See* **Equuleus**

Winged Horse. *See* **Pegasus**

Horseshoe (bright) Nebula in Sagittarius (Checkmark Nebula, M17, NGC 6618, Omega Nebula, Swan Nebula), *18h 20.8m, –16° 11'*

Hounds. *See* **Canes Venatici**

Hourglass (bright) Nebula in Sagittarius, *18h 03.8m, –24° 23'*

Hubble's Variable (bright) Nebula in Monoceros (NGC 2261), *6h 39.2m, +8° 44'*

Hunter. *See* **Orion**

Hunting Dogs. *See* **Canes Venatici**

Huruhata's Object binary star in Canis Minor, *7h 30.3m, +4° 32'*

Hya. *See* **Hydra**

Hyades open cluster in Taurus (Melotte 25), *4h 26.9m, +15° 52'*

Hyadum I (gamma (γ) Tauri), *4h 19.8m, +15° 38'*

Hyadum II (delta^1 (δ^1) Tauri), *4h 22.9m, +17° 33'*

Hydra (abbreviated **Hya**, Sea Serpent, Water Monster):

cluster, globular, M68 (NGC 4590), *12h 39.5m, –26° 45'*

cluster, open, M48 (NGC 2548), *8h 13.8m, –5° 48'*

galaxies:

Hydra A, *9h 18.1m, –12° 06'*

Hydra I cluster, *10h 36.9m, –27° 32'*

Hydra II cluster, *8h 58.0m, +3° 09'*

M83 (NGC 5236), *13h 37.0m, –29° 52'*

nebula, planetary, Ghost of Jupiter (Jupiter Nebula, NGC 3242), *10h 24.8m, –18° 38'*

star, alpha (a) Hydrae (Alfard, Alphard, Cor Hydrae, Red Bird), *9h 27.6m, –8° 40'*

Hydra A Galaxy, *9h 18.1m, –12° 06'*

Hydra I cluster of galaxies, *10h 36.9m, –27° 32'*

Hydra II cluster of galaxies, *8h 58.0m, +3° 09'*

Hydrus (abbreviated **Hyi**, Water Snake)

i Bootis (44 Bootis), *15h 03.8m, +47° 39'*

IC 59 and 63 bright nebula in Cassiopeia (gamma (γ) Cassiopeiae Nebula), *0h 56.7m, +61° 04'* and *0h 59.5m, +60° 49'*

IC 405 bright nebula in Auriga (Flaming Star Nebula, Schaeberle's Flaming Star), *5h 16.2m, +34° 16'*

IC 708 galaxy in Ursa Major (Papillon), *11h 33.9m, +49° 03'*

IC 1318 bright nebula in Cygnus (Cygnus Nebula, gamma (γ) Cygni Nebula), *20h 22.2m, +40° 15'*

IC 2118 bright nebula in Eridanus (Witch Head Nebula), *5h 06.9m, –7° 13'*

IC 2177 bright nebula in Monoceros (Eagle Nebula), *7h 05.1m, –10° 42'*

IC 2220 bright nebula in Carina (Butterfly Nebula, Toby Jug Nebula), *7h 56.8m, –59° 07'*

IC 2391 open cluster in Vela (omicron (o) Velorum Cluster, Velorum), *8h 40.2m, –53° 04'*

IC 2574 galaxy in Ursa Major (Coddington's "Nebula"), *10h 28.4m, +68° 25'*

IC 2602 open cluster in Carina (Southern Pleiades, theta (θ) Carinae), *10h 43.2m, –64° 24'*

IC 2944 bright nebula in Centaurus (Running Chicken Nebula), *11h 36.6m, –63° 02'*

IC 2944 and 2948 bright nebula in Centaurus (lambda (λ) Centauri Nebula), *11h 36.6m, –63° 02'* and *11h 38.8m, –63° 32'*

IC 4604 (bright) rho (ρ) Ophiuchi Nebula, *16h 25.6m, –23° 26'*

IC 4725 open cluster in Sagittarius (M25), *18h 31.6m, –19° 15'*

IC 5067 and 5070 bright nebula in Cygnus (Pelican Nebula), *20h 47.8m, +44° 22'* and *20h 50.8m, +44° 21'*

IC 5146 bright nebula in Cygnus (Cocoon Nebula), *21h 53.4m, +47° 16'*

Indus (abbreviated **Ind**, Indian)

Inne's Star (Proxima Centauri), *14h 29.7m, –62° 41'*

Integral Sign Galaxy in Camelopardalis, *7h 11.4m, +71° 50'*

Intergalactic Wanderer globular cluster in Lynx (NGC 2419), *7h 38.1m, +38° 53'*

iota (ι) Carinae (Aspidiske), *9h 17.1m, –59° 17'*

iota (ι) Ceti (Deneb Kaitos, Schemali), *0h 19.4m, –8° 49'*

iota (ι) Draconis (Edasich), *15h 24.9m, +58° 58'*

iota (ι) Ursae Majoris (Dnoces, Talita, Talitha), *8h 59.2m, +48° 02'*

iota (ι) Virginis (Syrma), *14h 16.0m, –6° 00'*

Iron Star in Ophiuchus, *17h 43.9m, –6° 16'*

Izar (epsilon (ε) Bootis, Mizar, Pulcherrima), *14h 45.0m, +27° 04'*

Jj

Jabbah (nu (ν) Scorpii), *16h 12.0m, –19°27'*

Jewel Box open cluster in Crux (kappa (κ) Crucis Cluster, NGC 4755), *12h 53.6m, –60°20'*

Job's Coffin. *See* **Delphinus**

Jupiter (planetary) Nebula in Hydra (Ghost of Jupiter, NGC 3242), *10h 24.8m, –18°38'*

Kk

Kaitain (alpha (α) Piscium, Alrescha, Alrischa, Okda), *2h 02.0m, +2° 46'*

kappa (κ) Aquarii (Situla), *22h 37.8m, –4° 14'*

kappa (κ) Crucis (open) Cluster (Jewel Box, NGC 4755), *12h 53.6m, –60° 20'*

kappa (κ) Herculis (Marfak, Marfik), *16h 08.1m, +17° 03'*

kappa (κ) Orionis (Saiph), *5h 47.8m, –9° 40'*

Kapteyn's Star in Pictor, *5h 10.6m, –44° 52'*

Kaus Australis (epsilon (ε) Sagittarii), *18h 24.2m, –34° 23'*

Kaus Borealis (lambda (λ) Sagittarii), *18h 28.0m, –25° 25'*

Kaus Meridionalis (delta (δ) Sagittarii, Media), *18h 21.0m, –29° 50'*

Keel of Argo Navis. *See* **Carina**

Keenan's System of galaxies in Ursa Major (NGC 5216 and 5218), *13h 32.1m, +62° 42'* and *13h 32.2m, +62° 46'*

Keid (omicron^2 (o^2) Eridani), *4h 15.3m, –7° 39'*

Kemble's Cascade star chain in Camelopardalis, *3h 57.0m, +63° 00'*

Kepler's Star in Ophiuchus, *17h 30.6m, –21° 29'*

Keyhole (dark) Nebula in Carina, *10h 43.8m, –59° 52'*

Keystone. *See* **Hercules**

Kids (epsilon, eta, and zeta (ε, η, and ζ) Aurigae), *5h 02.0m, +43° 49'; 5h 06.5m, +41° 14';* and *5h 02.5m, +41° 05'*

Kiffa Australis (alpha^2 (α^2) Librae, Zubenelgenubi), *14h 50.9m, –16° 02'*

Kiffa Borealis (beta (β) Librae, Zubeneschamali), *15h 17.0m, –9° 23'*

King. *See* **Cepheus**

Kitalpha (alpha (α) Equulei), *21h 15.8m, +5° 15'*

Kite. *See* **Auriga**

Klemola's Star in Leo, *10h 38.9m, +10°04'*
Kneeler. *See* **Hercules**
Kochab (beta (β) Ursae Minoris), *14h 50.7m, +74°09'*
Kornephoros (beta (β) Herculis), *16h 30.2m, +21°29'*
Krzeminski's Star in Centaurus, *11h 21.3m, –60°37'*
Kurhah (xi (ξ) Cephei), *22h 03.8m, +64°38'*
Kursa (beta (β) Eridani, Cursa, Dhalim), *5h 07.8m, –5°05'*
Kurtz's Light Variable star in Octans, *20h 03.6m, –78°50'*
Kutner's Cloud dark nebula in Taurus, *4h 33.0m, +24°36'*
Kuwano's Object star in Vulpecula, *20h 21.2m, +21°36'*

l Carinae, *9h 45.2m, –62° 30'*

L¹ Puppis, *7h 13.2m, –45° 11'*

L² Puppis, *7h 13.5m, –44° 38'*

La Superba (Y Canum Venaticorum), *12h 45.1m, +45° 26'*

Lac. *See* **Lacerta**

Lacaille 8760 star in Microscopium, *21h 17.3m, –38° 52'*

Lacaille 9352 star in Piscis Austrinus, *23h 05.9m, –35° 51'*

Lacerta (abbreviated **Lac**, Lizard):

    star, Babcock's Magnetic Star, *22h 44.2m, +55° 35'*

Lacework (bright) Nebula in Cygnus (Filamentary Nebula, NGC 6960, part of Veil Nebula), *20h 45.7m, +30° 43'*

Lady in the Chair. *See* **Cassiopeia**

Lagoon (bright) Nebula in Sagittarius (M8, NGC 6523), *18h 03.8m, –24° 23'*

Lalande 21185 star in Ursa Major, *11h 03.3m, +35° 58'*

lambda (λ) Centauri (bright) Nebula (IC 2944 and 2948), *11h 36.6m, –63° 02'* and *11h 38.8m, –63° 32'*

lambda (λ) Draconis (Giansar or Giauzar), *11h 31.4m, +69° 20'*

lambda (λ) Herculis (Masym), *17h 30.7m, +26° 07'*

lambda (λ) Leonis (Alterf), *9h 31.7m, +22° 58'*

lambda (λ) Ophiuchi (Marfak, Marfik), *16h 30.9m, +1° 59'*

lambda (λ) Orionis (Meissa), *5h 35.1m, +9° 56'*

lambda (λ) Orionis (open) Cluster and (bright) Nebula, *5h 35.1m, +9° 56'*

lambda and mu (λ and μ) Pegasi (Sadalbari), *22h 46.5m, +23° 34'* and *22h 50.0m, +24° 36'*

lambda (λ) Sagittarii (Kaus Borealis), *18h 28.0m, –25° 25'*

lambda (λ) Scorpii (Shaula), *17h 33.6m, –37° 06'*

lambda (λ) Ursae Majoris (Tania Borealis), *10h 17.1m, +42° 55'* lambda (λ) Velorum (Suhail), *9h 08.0m, –43° 26'*

Lamont's Star in Andromeda, *0h 45.4m, +41° 32'*

Large Bear. *See* **Ursa Major**

Large Dog. *See* **Canis Major**

Large Magellanic Cloud galaxy in Dorado and Mensa (Great Magellanic Cloud, LMC, Nubecula Major), centered in Dorado at *5h 23.6m, –69° 45'*

Leo (abbreviated **Leo**, Lion, Sickle):

galaxies:

Copeland's Septet (NGC 3745, 3746, 3748, 3750, 3751, 3753, and 3754), *11h 37.7m, +22° 01'; 11h 37.7m, +22° 00'; 11h 37.8m, +22° 02'; 11h 37.9m, +21° 58'; 11h 37.9m, +21° 56'; 11h 37.9m, +21° 59';* and *11h 37.9m, +21° 59'*

Leo I (Regulus Dwarf Galaxy), *10h 08.4m, +12° 18'*

Leo II (Leo B), *11h 15.5m, +22° 10'*

Leo III (Leo A), *9h 59.4m, +30° 45'*

Leo A (Leo III), *9h 59.4m, +30° 45'*

Leo A Cluster, *11h 10.9m, +28° 41'*

Leo B (Leo II), *11h 15.5m, +22° 10'*

Leo Cluster, *10h 27.8m, +10° 25'*

Leo Triplet (M65, M66, and NGC 3628), *11h 18.9m, +13° 05'; 10h 46.8m, +12° 59';* and *11h 20.3m, +13° 36'*

M65 (NGC 3623, pair with M66), *11h 18.9m, +13° 05'*

M66 (NGC 3627, pair with M65), *11h 20.3m, +12° 59'*

M95 (NGC 3351, pair with M96), *10h 44.0m, +11° 42'*

M96 (NGC 3368, pair with M95), *10h 46.8m, +11° 49'*

M105 (NGC 3379), *10h 47.8m, +12° 35'*

Regulus Dwarf Galaxy (Leo I), *10h 08.4m, +12° 18'*

stars:

alpha (α) Leonis (Cor Leonis, Regulus), *10h 08.4m, +11° 58'*

beta (β) Leonis (Dafira, Deneb, Denebola, Serpha), *11h 49.1m, +14° 34'*

Leo (cont.):

    stars (cont.):

        delta (δ) Leonis (Dhur, Duhr, Zosma, Zubra), *11h 14.1m, +20°31'*

        epsilon (ε) Leonis (Algenubi), *9h 45.8m, +23°46'*

        gamma^1 (γ^1) Leonis (Algeiba, Algieba), *10h 20.0m, +19°50'*

        Klemola's Star, *10h 38.9m, +10°04'*

        lambda (λ) Leonis (Alterf), *9h 31.7m, +22°58'*

        mu (μ) Leonis (Alshemali, Rasalas), *9h 52.8m, +26°00'*

        Scheiner's Star, *9h 37.2m, +15°14'*

        theta (θ) Leonis (Chertan, Chort), *11h 14.2m, +15°26'*

        zeta (ζ) Leonis (Adhafera, Aldhafera), *10h 16.7m, +23°25'*

Leo A Cluster of galaxies, *11h 10.9m, +28°41'*

Leo A galaxy (Leo III), *9h 59.4m, +30°45'*

Leo B galaxy (Leo II), *11h 15.5m, +22°10'*

Leo I galaxy (Regulus Dwarf Galaxy), *10h 08.4m, +12°18'*

Leo II galaxy (Leo B), *11h 15.5m, +22°10'*

Leo III galaxy (Leo A), *9h 59.4m, +30°45'*

Leo Cluster of galaxies, *10h 27.8m, +10°25'*

Leo Minor (abbreviated **LMi**, Lesser Lion):

    galaxy, Spider, *10h 42.7m, +34°26'*

Leo Triplet of galaxies (M65, M66, and NGC 3628), *11h 18.9m, +13°05'; 10h 46.8m, +12°59';* and *11h 20.3m, +13°36'*

Lepus (abbreviated **Lep**, Hare):

    cluster, globular, M79 (NGC 1904), *5h 24.5m, –24°33'*

    stars:

        alpha (α) Leporis (Arneb, Arsh), *5h 32.7m, –17°49'*

        beta (β) Leporis (Nihal), *5h 28.2m, –20°46'*

        R Leporis (Crimson Star, Hind's Crimson Star), *4h 59.6m, –14°48'*

Lesath (Lesuth, upsilon (υ) Scorpii), *17h 30.8m, –37°18'*

Lesser Bear. *See* **Ursa Minor**

Lesser Dog. *See* **Canis Minor**

Lesser Lion. *See* **Leo Minor**

Lesser Magellanic Cloud galaxy in Tucana (NGC 292, Nubecula
Minor, Small Magellanic Cloud, SMC), *0h 52.7m, –72°50'*

Lesuth (Lesath, upsilon (υ) Scorpii), *17h 30.8m, –37°18'*

Level. *See* **Norma**

Libra (abbreviated **Lib**, Balance, Scales):
    galaxy, Fath 703, *15h 13.8m, –15°28'*
    stars:
        alpha^2 (α^2) Librae (Kiffa Australis, Zubenelgenubi),
            14h 50.9m, –16°02'
        beta (β) Librae (Kiffa Borealis, Zubeneschamali),
            15h 17.0m, –9°23'
        33 G. Librae, *14h 57.5m, –21°25'*

Liller's Star in Centaurus, *11h 21.2m, –60°36'*

Lindsay-Shapley Ring galaxy in Volans, *6h 43.1m, –74°14'*

Lions. *See* **Leo; Leo Minor**

Little Bear. *See* **Ursa Minor**

Little Dipper. *See* **Ursa Minor**

Little Dog. *See* **Canis Minor**

Little Dog Star (alpha (α) Canis Minoris, Procyon), *7h 39.3m,
+5°14'*

Little Dumbbell (planetary) Nebula in Perseus (Butterfly Nebula,
Cork Nebula, M76, NGC 650 and 651), *1h 42.3m, +51°34'*
and *1h 42.3m, +51°35'*

Little Fox. *See* **Vulpecula**

Little Gem planetary nebula in Sagittarius (NGC 6445),
17h 49.2m, –20°01'

Little Horse. *See* **Equuleus**

Little Lion. *See* **Leo Minor**

Lizard. *See* **Lacerta**

LMC galaxy in Dorado and Mensa (Great Magellanic Cloud, Large
Magellanic Cloud, Nubecula Major), centered in Dorado at
5h 23.6m, –69°45'

LMi. *See* **Leo Minor**

Loop (bright) Nebula in Dorado (Great Looped Nebula, NGC 2070, Tarantula Nebula, 30 Doradus Nebula), *5h 38.7m, –69°06'*

Lord Rosse's "Nebula" galaxy in Canes Venatici (M51, NGC 5194 and 5195, Question Mark Galaxy, Whirlpool Galaxy), *13h 29.9m, +47°12' and 13h 30.0m, +47°16'*

Lower's (bright) Nebula in Orion, *6h 08.9m, +15°49'*

Lupus (abbreviated **Lup**, Wolf)

Luyten's Flare Star in Cetus, *1h 38.8m, –17°58'*

Luyten's Star in Aquila, *19h 42.1m, +6°00'*

Lynx (abbreviated **Lyn**, Lynx, Tiger):

    cluster, globular, Intergalactic Wanderer (NGC 2419), *7h 38.1m, +38°53'*

    galaxy, Bear Claw Galaxy (Bear Paw Galaxy, NGC 2537), *8h 13.2m, +46°00'*

Lyra (abbreviated **Lyr**, Harp, Lyre):

    cluster, globular, M56 (NGC 6779), *19h 16.6m, +30°11'*

    cluster, open, delta (δ) Lyrae Cluster, *18h 53.5m, +36°55'*

    nebula, planetary, M57 (NGC 6720, Ring Nebula), *18h 53.6m, +33°02'*

    stars:

        alpha (α) Lyrae (Harp Star, Vega, Wega), *18h 36.9m, +38°47'*

        alpha (α) Lyrae, alpha (α) Aquilae, and alpha (α) Cygni (Summer Triangle), *18h 36.9m, +38°47'; 19h 50.8m, +8°52'; and 20h 41.4m, +45°17'*

        beta (β) Lyrae (Sheliak, Shelyak), *18h 50.1m, +33°22'*

        epsilon[1] and epsilon[2] (ε^1 and ε^2) Lyrae (Double-Double), *18h 44.3m, +39°40' and 18h 44.4m, +39°37'*

        gamma (γ) Lyrae (Sulafat, Sulaphat), *18h 58.9m, +32°41'*

Lyre. *See* **Lyra**

M1 bright nebula in Taurus (Crab Nebula, NGC 1952), *5h 34.5m, +22°01'*

M2 globular cluster in Aquarius (NGC 7089), *21h 33.5m, –0°49'*

M3 globular cluster in Canes Venatici (NGC 5272), *13h 42.2m, +28°23'*

M4 globular cluster in Scorpius (NGC 6121), *16h 23.6m, –26°32'*

M5 globular cluster in Serpens (NGC 5904), *15h 18.6m, +2°05'*

M6 open cluster in Scorpius (Butterfly Cluster, NGC 6405), *17h 40.1m, –32°13'*

M7 open cluster in Scorpius (NGC 6475), *17h 53.9m, –34°49'*

M8 bright nebula in Sagittarius (Lagoon Nebula, NGC 6523), *18h 03.8m, –24°23'*

M9 globular cluster in Ophiuchus (NGC 6333), *17h 19.2m, –18°31'*

M10 globular cluster in Ophiuchus (NGC 6254), *16h 57.1m, –4°06'*

M11 open cluster in Scutum (NGC 6705, Wild Duck Cluster), *18h 51.1m, –6°16'*

M12 globular cluster in Ophiuchus (NGC 6218), *16h 47.2m, –1°57'*

M13 globular cluster in Hercules (Great Cluster in Hercules, Hercules Cluster, NGC 6205), *16h 41.7m, +36°28'*

M14 globular cluster in Ophiuchus (NGC 6402), *17h 37.6m, –3°15'*

M15 globular cluster in Pegasus (NGC 7078), *21h 30.0m, +12°10'*

M16 bright nebula in Serpens (Eagle Nebula, NGC 6611, Star Queen Nebula), *18h 18.8m, –13°47'*

M17 bright nebula in Sagittarius (Checkmark Nebula, Horseshoe Nebula, NGC 6618, Omega Nebula, Swan Nebula), *18h 20.8m, –16°11'*

M18 open cluster in Sagittarius (NGC 6613), *18h 19.9m, –17°08'*

M19 globular cluster in Ophiuchus (NGC 6273), *17h 02.6m, –26°16'*

M20 bright nebula in Sagittarius (NGC 6514, Trifid Nebula),
18h 02.3m, –23°02'

M21 open cluster in Sagittarius (NGC 6531), *18h 04.6m, –22°30'*

M22 globular cluster in Sagittarius (NGC 6656), *18h 36.4m,
–23°54'*

M23 open cluster in Sagittarius (NGC 6494), *17h 56.8m, –19°01'*

M24 star cloud in Sagittarius (Delle Caustiche, Small Sagittarius
Cloud), *18h 16.9m, –18°29'*

M25 open cluster in Sagittarius (IC 4725), *18h 31.6m, –19°15'*

M26 open cluster in Scutum (NGC 6694), *18h 45.2m, –9°24'*

M27 planetary nebula in Vulpecula (Diabolo Nebula,
Double-Headed Shot, Dumbbell Nebula, NGC 6853),
19h 59.6m, +22°43'

M28 globular cluster in Sagittarius (NGC 6626), *18h 24.5m,
–24°52'*

M29 open cluster in Cygnus (NGC 6913), *20h 23.9m, +38°32'*

M30 globular cluster in Capricornus (NGC 7099), *21h 40.4m,
–23°11'*

M31 galaxy in Andromeda (Andromeda Galaxy, Great Galaxy in
Andromeda, NGC 224), *0h 42.7m, +41°16'*

M32 galaxy in Andromeda (NGC 221), *0h 42.7m, +40°52'*

M33 galaxy in Triangulum (NGC 598, Pinwheel Galaxy,
Triangulum Galaxy), *1h 33.9m, +30°39'*

M34 open cluster in Perseus (NGC 1039), *2h 42.0m, +42°47'*

M35 open cluster in Gemini (NGC 2168), *6h 08.9m, +24°20'*

M36 open cluster in Auriga (NGC 1960), *5h 36.1m, +34°08'*

M37 open cluster in Auriga (NGC 2099), *5h 52.4m, +32°33'*

M38 open cluster in Auriga (NGC 1912), *5h 28.7m, +35°50'*

M39 open cluster in Cygnus (NGC 7092), *21h 32.2m, +48°26'*

M40 double star in Ursa Major, *12h 22.2m, +58°05'*

M41 open cluster in Canis Major (NGC 2287), *6h 47.0m, –20°44'*

M42 bright nebula in Orion (Great Nebula in Orion, NGC 1976,
Orion Nebula), *5h 35.4m, –5°27'*

M43 detached portion of M42 (NGC 1982), *5h 35.6m, –5°16'*

M44 open cluster in Cancer (Beehive Cluster, Manger, NGC 2632, Praesepe Cluster), *8h 40.1m, +19°59'*

M45 open cluster in Taurus (Pleiades, Seven Sisters), *3h 47.0m, +24°07'*

M46 open cluster in Puppis (NGC 2437), *7h 41.8m, –14°49'*

M47 open cluster in Puppis (NGC 2422), *7h 36.6m, –14°30'*

M48 open cluster in Hydra (NGC 2548), *8h 13.8m, –5°48'*

M49 galaxy in Virgo (NGC 4472), *12h 29.8m, +8°00'*

M50 open cluster in Monoceros (NGC 2323), *7h 03.2m, –8°20'*

M51 galaxy in Canes Venatici (Lord Rosse's "Nebula," NGC 5194 and 5195, Question Mark Galaxy, Whirlpool Galaxy), *13h 29.9m, +47°12' and 13h 30.0m, +47°16'*

M52 open cluster in Cassiopeia (NGC 7654), *23h 24.2m, +61°35'*

M53 globular cluster in Coma Berenices (NGC 5024), *13h 12.9m, +18°10'*

M54 globular cluster in Sagittarius (NGC 6715), *18h 55.1m, –30°29'*

M55 globular cluster in Sagittarius (NGC 6809), *19h 40.0m, –30°58'*

M56 globular cluster in Lyra (NGC 6779), *19h 16.6m, +30°11'*

M57 planetary nebula in Lyra (NGC 6720, Ring Nebula), *18h 53.6m, +33°02'*

M58 galaxy in Virgo (NGC 4579), *12h 37.7m, +11°49'*

M59 galaxy in Virgo (NGC 4621), *12h 42.0m, +11°39'*

M60 galaxy in Virgo (NGC 4649), *12h 43.7m, +11°33'*

M61 galaxy in Virgo (NGC 4303), *12h 21.9m, +4°28'*

M62 globular cluster in Ophiuchus and Scorpius (NGC 6266), *17h 01.2m, –30°07'*

M63 galaxy in Canes Venatici (NGC 5055, Sunflower Galaxy), *13h 15.8m, +42°02'*

M64 galaxy in Coma Berenices (Blackeye Galaxy, NGC 4826), *12h 56.7m, +21°41'*

M65 galaxy in Leo (NGC 3623, pair with M66), *11h 18.9m, +13°05'*

M66 galaxy in Leo (NGC 3627, pair with M65), *11h 20.3m, +12°59'*

M67 open cluster in Cancer (NGC 2682), *8h 50.4m, +11° 49'*

M68 globular cluster in Hydra (NGC 4590), *12h 39.5m, –26° 45'*

M69 globular cluster in Sagittarius (NGC 6637), *18h 31.4m, –32° 21'*

M70 globular cluster in Sagittarius (NGC 6681), *18h 43.2m, –32° 18'*

M71 globular cluster in Sagitta (NGC 6838), *19h 53.8m, +18° 47'*

M72 globular cluster in Aquarius (NGC 6981), *20h 53.5m, –12° 32'*

M73 open cluster in Aquarius (NGC 6994), *20h 59.0m, –12° 38'*

M74 galaxy in Pisces (NGC 628), *1h 36.7m, +15° 47'*

M75 globular cluster in Sagittarius (NGC 6864), *20h 06.1m, –21° 55'*

M76 planetary nebula in Perseus (Butterfly Nebula, Cork Nebula, Little Dumbbell Nebula, NGC 650 and 651), *1h 42.3m, +51° 34'* and *1h 42.3m, +51° 35'*

M77 galaxy in Cetus (NGC 1068), *2h 42.7m, –0° 01'*

M78 bright nebula in Orion (NGC 2068), *5h 46.7m, +0° 03'*

M79 globular cluster in Lepus (NGC 1904), *5h 24.5m, –24° 33'*

M80 globular cluster in Scorpius (NGC 6093), *16h 17.0m, –22° 59'*

M81 Dwarf B Galaxy in Ursa Major, *10h 05.5m, +70° 22'*

M81 galaxy in Ursa Major (Bode's "Nebulae" with NGC 3034, NGC 3031), *9h 55.6m, +69° 04'*

M81 Group cluster of galaxies in Ursa Major, *10h 02.1m, +68° 45'*

M82 galaxy in Ursa Major (Bode's "Nebulae" with NGC 3031, NGC 3034, member of M81 group), *9h 55.8m, +69° 41'*

M83 galaxy in Hydra (NGC 5236), *13h 37.0m, –29° 52'*

M84 galaxy in Virgo (NGC 4374), *12h 25.1m, +12° 53'*

M85 galaxy in Coma Berenices (NGC 4382), *12h 25.4m, +18° 11'*

M86 galaxy in Virgo (NGC 4406), *12h 26.2m, +12° 57'*

M87 galaxy in Virgo (NGC 4486, Virgo A Galaxy), *12h 30.8m, +12° 24'*

M88 galaxy in Coma Berenices (NGC 4501), *12h 32.0m, +14° 25'*

M89 galaxy in Virgo (NGC 4552), *12h 35.7m, +12° 33'*

M90 galaxy in Virgo (NGC 4569), *12h 36.8m, +13° 10'*

M91—error in Messier Catalogue or possibly galaxy in Coma Bernenices (NGC 4548), *12h 35.4m, +14° 30'*

M92 globular cluster in Hercules (NGC 6341), *17h 17.1m, +43° 08'*

M93 open cluster in Puppis (NGC 2447), *7h 44.6m, –23° 52'*

M94 galaxy in Canes Venatici (NGC 4736), *12h 50.9m, +41° 07'*

M95 galaxy in Leo (NGC 3351, pair with M96), *10h 44.0m, +11° 42'*

M96 galaxy in Leo (NGC 3368, pair with M95), *10h 46.8m, +11° 49'*

M97 planetary nebula in Ursa Major (NGC 3587, Owl Nebula), *11h 14.8m, +55° 01'*

M98 galaxy in Coma Berenices (NGC 4192), *12h 13.8m, +14° 54'*

M99 galaxy in Coma Berenices (NGC 4254, Pinwheel Galaxy), *12h 18.8m, +14° 25'*

M100 galaxy in Coma Berenices (NGC 4321), *12h 22.9m, +15° 49'*

M101 galaxy in Ursa Major (NGC 5457, Pinwheel Galaxy), *14h 03.2m, +54° 21'*

M101 Group cluster of galaxies in Ursa Major, *14h 09.4m, +54° 55'*

M102—accidental reobservation of M101

M103 open cluster in Cassiopeia (NGC 581), *1h 33.2m, +60° 42'*

M104 galaxy in Virgo (Darklane Galaxy, NGC 4594, Sombrero Galaxy), *12h 40.0m, –11° 37'*

M105 galaxy in Leo (NGC 3379), *10h 47.8m, +12° 35'*

M106 galaxy in Canes Venatici (NGC 4258), *12h 19.0m, +47° 18'*

M107 globular cluster in Ophiuchus (NGC 6171), *16h 32.5m, –13° 03'*

M108 galaxy in Ursa Major (NGC 3556), *11h 11.5m, +55° 40'*

M109 galaxy in Ursa Major (NGC 3992), *11h 57.6m, +53° 23'*

M110 galaxy in Andromeda (NGC 205), *0h 40.4m, +41° 41'*

Maffei I galaxy in Cassiopeia, *2h 36.3m, +59° 39'*

Maffei II galaxy in Cassiopeia, *2h 41.9m, +59° 36'*

Magellanic Clouds:

Large Magellanic Cloud galaxy in Dorado and Mensa (Great Magellanic Cloud, LMC, Nubecula Major), centered in Dorado at *5h 23.6m, –69° 45'*

Magellanic Clouds (cont.):

    Small Magellanic Cloud galaxy in Tucana (Lesser Magellanic Cloud, NGC 292, Nubecula Minor, SMC), *0h 52.7m, –72°50'*

Maia (20 Tauri), *3h 45.8m, +24°22'*

Maia (bright) Nebula in Taurus (NGC 1432), *3h 45.8m, +24°22'*

Malus (Mast of Argo Navis). *See* **Pyxis**

Manger open cluster in Cancer (Beehive Cluster, M44, NGC 2632, Praesepe Cluster), *8h 40.1m, +19°59'*

Marchab (alpha (α) Pegasi, Markab) *23h 04.8m, +15°12'*

Marfak or Marfik:

    alpha (α) Persei (Alchemb, Algenib, Mirfak), *3h 24.3m, +49°52'*

    kappa (κ) Herculis, *16h 08.1m, +17°03'*

    lambda (λ) Ophiuchi, *16h 30.9m, +1°59'*

    mu (μ) Cassiopeiae, *1h 08.3m, +54°55'*

    theta (θ) Cassiopeiae, *1h 11.1m, +55°09'*

Mariner's Compass, Compass of Argo Navis (**Pyxis**, abbreviated **Pyx**)

Markab (alpha (α) Pegasi, Marchab), *23h 04.8m, +15°12'*

Martial Star (alpha (α) Orionis, Betelgeuse, Mirzam), *5h 55.2m, +7°24'*

Mast of Argo Navis (Malus). *See* **Pyxis**

Masym (lambda (λ) Herculis), *17h 30.7m, +26°07'*

Matar (eta (η) Pegasi, Sadalmatar), *22h 43.0m, +30°13'*

Mayall's Object galaxy in Ursa Major, *11h 03.9m, +40°51'*

McLeish's Object galaxy in Pavo, *20h 09.7m, –66°13'*

Meboula (epsilon (ε) Geminorum, Mebsuta, Mebusta), *6h 43.9m, +25°08'*

Media (delta (δ) Sagittarii, Kaus Meridionalis), *18h 21.0m, –29°50'*

Medusa (planetary) Nebula in Gemini, *7h 29.0m, +13°15'*

Medusa Star (Algol, beta (β) Persei, Demon Star), *3h 08.2m, +40°57'*

Megrez (delta (δ) Ursae Majoris), *12h 15.4m, +57°02'*

Meissa (lambda (λ) Orionis), *5h 35.1m, +9°56'*

Mekbuda (zeta (ζ) Geminorum), *7h 04.1m, +20° 34'*

Melotte 25 open cluster in Taurus (Hyades), *4h 26.9m, +15° 52'*

Melotte 111 open cluster in Coma Berenices (Coma Star Cluster),
12h 25.0m, +26° 00'

Men. *See* **Mensa**

Menkab (alpha (α) Ceti, Menkar), *3h 02.3m, +4° 05'*

Menkalinan (beta (β) Aurigae), *5h 59.5m, +44° 57'*

Menkar (alpha (α) Ceti, Menkab), *3h 02.3m, +4° 05'*

Menkent (theta (θ) Centauri), *14h 06.7m, –36° 22'*

Menkib (xi (ξ) Persei), *3h 59.0m, +35° 47'*

Mensa (abbreviated **Men**, Table Mountain):

galaxy, Large Magellanic Cloud (Great Magellanic Cloud,
LMC, Nubecula Major), centered in Dorado at
5h 23.6m, –69° 45'

star, Sanduleak's Star, *5h 45.3m, –71° 16'*

Merak (beta (β) Ursae Majoris, Mirak), *11h 01.8m, +56° 23'*

Merope (23 Tauri), *3h 46.3m, +23° 57'*

Merope (bright) Nebula in Taurus (NGC 1435), *3h 46.1m, +23° 47'*

Merrill's Star in Sagitta, *19h 11.5m, +16° 52'*

Mesartim (gamma^2 (γ^2) Arietis), *1h 53.5m, +19° 18'*

Miaplacidus (beta (β) Carinae), *9h 13.2m, –69° 43'*

Mic. *See* **Microscopium**

Mice galaxies in Coma Berenices (NGC 4676), *12h 46.1m, +30° 44'*

Microscopium (abbreviated **Mic**, Microscope):

nebula, dark, Hoffmeister's Cloud, *20h 47.0m, –42° 00'*

star, Lacaille 8760, *21h 17.3m, –38° 52'*

Milk Dipper. *See* **Sagittarius**

Milky Way galaxy centered in Sagittarius at *17h 45.6m, –28° 56'*

Milky Way star cloud in Sagittarius (M24, encloses NGC 6603),
18h 18.4m, –18° 25'

Mimosa (beta (β) Crucis), *12h 47.7m, –59° 41'*

Miniature Spiral Galaxy in Ursa Major (NGC 3928), *11h 51.8m,
+48° 41'*

Minkowski's Footprint bright nebula in Cygnus (Footprint
  Nebula), *19h 36.3m, +29° 33'*

Minkowski's Object galaxy in Cetus, *1h 25.8m, –1° 22'*

Mintaka (delta (δ) Orionis), *5h 32.0m, –0° 18'*

Mira (omicron (o) Ceti, Wonderful Star), *2h 19.3m, –2° 59'*

Mirach (beta (β) Andromedae), *1h 09.7m, +35° 37'*

Mirak (beta (β) Ursae Majoris, Merak), *11h 01.8m, +56° 23'*

Mirfak (Alchemb, Algenib, alpha (α) Persei, Marfak, Marfik),
  3h 24.3m, +49° 52'

Mirzam:

    alpha (α) Orionis (Betelgeuse, Martial Star), *5h 55.2m, +7° 24'*

    beta (β) Canis Majoris (Murzim), *6h 22.7m, –17° 57'*

Mizar:

    epsilon (ε) Bootis (Izar, Pulcherrima), *14h 45.0m, +27° 04'*

    zeta (ζ) Ursae Majoris, *13h 23.9m, +54° 56'*

Mon. *See* **Monoceros**

Monarch. *See* **Cepheus**

Monoceros (abbreviated **Mon**, Unicorn):

    clusters, open:

        Christmas Tree Cluster (in NGC 2264), *6h 41.1m, +9° 53'*

        M50 (NGC 2323), *7h 03.2m, –8° 20'*

    nebulae, bright:

        Eagle Nebula (IC 2177), *7 h 05.1, –10° 42'*

        Hubble's Variable Nebula (NGC 2261), *6h 39.2m, +8° 44'*

        Monoceros Loop, *6h 38.0m, +6° 27'*

        Rosette Nebula (NGC 2237, 2238, 2239, and 2246),
  *6h 30.3m, +5° 03'; 6h 30.6m, +5° 01'; 6h 31.0m,
  +4° 57'; and 6h 32.4m, +5° 07'*

    nebulae, dark, Cone Nebula (Conus Nebula, in NGC 2264),
  6h 40.9m, +9° 54'

    nebulae, planetary, Red Rectangle, *6h 20.0m, –10° 39'*

    stars:

        Brewer's Star, *6h 52.0m, –1° 39'*

        Plaskett's Star, *6h 37.5m, +6° 08'*

Monoceros Loop bright nebula, *6h 38.0m, +6°27'*

Moving (bright) Nebula in Pegasus, *0h 06.4m, +20°12'*

mu¹ (μ¹) Bootis (Alkalurops), *15h 24.5m, +37°23'*

mu (μ) Cassiopeiae (Marfak, Marfik), *1h 08.3m, +54°55'*

mu (μ) Cephei (Garnet Star, Herschel's Garnet Star), *21h 43.5m, +58°47'*

mu (μ) Draconis (Alrakis, Arrakis), *17h 05.3m, +54°28'*

mu (μ) Geminorum (Tejat), *6h 23.0m, +22°31'*

mu (μ) Leonis (Alshemali, Rasalas), *9h 52.8m, +26°00'*

mu (μ) Normae (open) Cluster (NGC 6169), *16h 34.1m, –44°03'*

mu (μ) Pegasi (Sadalbari), *22h 50.0m, +24°36'*

mu (μ) Ursae Majoris (Tania Australis), *10h 22.3m, +41°30'*

Muhlifain (gamma (γ) Centauri), *12h 41.5m, –48°58'*

Muliphen (gamma (γ) Canis Majoris), *7h 03.8m, –15°38'*

Muphrid (eta (η) Bootis), *13h 54.7m, +18°24'*

Murzim (beta (β) Canis Majoris, Mirzam), *6h 22.7m, –17°57'*

Musca (abbreviated **Mus**, Fly)

Muscida (omicron (o) Ursae Majoris), *8h 30.3m, +60°43'*

Nn

N Velorum, *9h 31.2m, –57°02'*

Nakkar (beta (β) Bootis, Nekkar), *15h 01.9m, +40°23'*

Naos (zeta (ζ) Puppis), *8h 03.6m, –40°00'*

Nashira (gamma (γ) Capricorni), *21h 40.1m, –16°40'*

Nath (Alnath, beta (β) Tauri, Elnath), *5h 26.3m, +28°36'*

Navi (gamma (γ) Cassiopeiae), *0h 56.7m, +60°43'*

Nekkar (beta (β) Bootis, Nakkar), *15h 01.9m, +40°23'*

Net, Rhomboidal Net (**Reticulum**, abbreviated **Ret**)

Network (bright) Nebula in Cygnus (NGC 6992 and 6995, part of Veil Nebula), *20h 56.4m, +31°43'* and *20h 57.1m, +31°13'*

NGC 104 globular cluster in Tucana (47 Tucanae), *0h 24.1m, –72°05'*

NGC 205 galaxy in Andromeda (M110), *0h 40.4m, +41°41'*

NGC 221 galaxy in Andromeda (M32), *0h 42.7m, +40°52'*

NGC 224 galaxy in Andromeda (Andromeda Galaxy, Great Galaxy in Andromeda, M31), *0h 42.7m, +41°16'*

NGC 253 galaxy in Sculptor (Sculptor Galaxy), *0h 47.6m, –25°17'*

NGC 292 galaxy in Tucana (Lesser Magellanic Cloud, Nubecula Minor, Small Magellanic Cloud, SMC), *0h 52.7m, –72°50'*

NGC 457 open cluster in Cassiopeia (Owl Cluster), *1h 19.1m, +58°20'*

NGC 581 open cluster in Cassiopeia (M103), *1h 33.2m, +60°42'*

NGC 598 galaxy in Triangulum (M33, Pinwheel Galaxy, Triangulum Galaxy), *1h 33.9m, +30°39'*

NGC 628 galaxy in Pisces (M74), *1h 36.7m, +15°47'*

NGC 650 and 651 planetary nebula in Perseus (Butterfly Nebula, Cork Nebula, Little Dumbbell Nebula, M76), *1h 42.3m, +51°34'* and *1h 42.3m, +51°35'*

NGC 869 and 884 open clusters in Perseus (Double Cluster,
Sword-Hand of Perseus), *2h 19.0m, +57°09'* and *2h 22.4m,
+57°07'*

NGC 1039 open cluster in Perseus (M34), *2h 42.0m, +42°47'*

NGC 1068 galaxy in Cetus (M77), *2h 42.7m, –0°01'*

NGC 1275 galaxy in Perseus (Perseus A), *3h 19.8m, +41°31'*

NGC 1316 galaxy in Fornax (Fornax A Galaxy), *3h 22.7m, –37°12'*

NGC 1432 bright nebula in Taurus (Maia Nebula), *3h 45.8m,
+24°22'*

NGC 1435 bright nebula in Taurus (Merope Nebula), *3h 46.1m,
+23°47'*

NGC 1499 bright nebula in Perseus (California Nebula), *4h 00.7m,
+36°37'*

NGC 1554 bright nebula in Taurus (Struve's Lost Nebula),
4h 21.8m, +19°32'

NGC 1555 bright nebula in Taurus (Hind's Variable Nebula),
4h 22.9m, +19°32'

NGC 1595 and 1598 galaxies in Caelum (Carafe Group, with
Carafe Galaxy), *4h 28.4m, –47°49'* and *4h 28.6m, –47°47'*

NGC 1904 globular cluster in Lepus (M79), *5h 24.5m, –24°33'*

NGC 1912 open cluster in Auriga (M38), *5h 28.7m, +35°50'*

NGC 1952 bright nebula in Taurus (Crab Nebula, M1), *5h 34.5m,
+22°01'*

NGC 1960 open cluster in Auriga (M36), *5h 36.1m, +34°08'*

NGC 1976 bright nebula in Orion (Great Nebula in Orion, M42,
Orion Nebula), *5h 35.4m, –5°27'*

NGC 1982 detached portion of Orion (bright) Nebula M42 (M43),
5h 35.6m, –5°16'

NGC 2032, 2033, 2034, and 2035 bright nebula in Dorado (Seagull
Nebula), *5h 35.3m, –67°34'; 5h 34.5m, –69°44'; 5h 35.7m,
–66°56';* and *5h 35.3m, –67°34'*

NGC 2068 bright nebula in Orion (M78), *5h 46.7m, +0°03'*

NGC 2070 bright nebula in Dorado (Great Looped Nebula, Loop
Nebula, Tarantula Nebula, 30 Doradus Nebula), *5h 38.7m,
–69°06'*

NGC 2099 open cluster in Auriga (M37), *5h 52.4m, +32°33'*

NGC 2168 open cluster in Gemini (M35), *6h 08.9m, +24°20'*

NGC 2237, 2238, 2239, and 2246 bright nebula in Monoceros (Rosette Nebula), *6h 30.3m, +5°03'; 6h 30.6m, +5°01'; 6h 31.0m, +4°57'; and 6h 32.4m, +5°07'*

NGC 2261 bright nebula in Monoceros (Hubble's Variable Nebula), *6h 39.2m, +8°44'*

NGC 2264 open cluster and dark nebula in Monoceros (Christmas Tree Cluster and Cone Nebula), *6h 41.0m, +9°53'*

NGC 2287 open cluster in Canis Major (M41), *6h 47.0m, –20°44'*

NGC 2323 open cluster in Monoceros (M50), *7h 03.2m, –8°20'*

NGC 2392 planetary nebula in Gemini (Clown Face Nebula, Eskimo Nebula), *7h 29.2m, +20°55'*

NGC 2419 globular cluster in Lynx (Intergalactic Wanderer), *7h 38.1m, +38°53'*

NGC 2422 open cluster in Puppis (M47), *7h 36.6m, –14°30'*

NGC 2437 open cluster in Puppis (M46), *7h 41.8m, –14°49'*

NGC 2447 open cluster in Puppis (M93), *7h 44.6m, –23°52'*

NGC 2537 galaxy in Lynx (Bear Claw Galaxy, Bear Paw Galaxy), *8h 13.2m, +46°00'*

NGC 2548 open cluster in Hydra (M48), *8h 13.8m, –5°48'*

NGC 2573 galaxy in Octans (Polarissima Australis), *1h 42.0m, –89°20'*

NGC 2632 open cluster in Cancer (Beehive Cluster, M44, Manger, Praesepe Cluster), *8h 40.1m, +19°59'*

NGC 2682 open cluster in Cancer (M67), *8h 50.4m, +11°49'*

NGC 2685 galaxy in Ursa Major (Helix Galaxy), *8h 55.6m, +58°44'*

NGC 3031 galaxy in Ursa Major (Bode's "Nebulae" with NGC 3034, M81), *9h 55.6m, +69°04'*

NGC 3034 galaxy in Ursa Major (Bode's "Nebulae" with NGC 3031, M82), *9h 55.8m, +69°41'*

NGC 3115 galaxy in Sextans (Spindle "Nebula"), *10h 05.2m, –7°43'*

NGC 3132 planetary nebula in Vela (Eight-burst Nebula), *10h 07.0', –40°26'*

NGC 3172 galaxy in Ursa Minor (Polarissima Borealis), *11h 50.0m, +89°07'*

NGC 3242 planetary nebula in Hydra (Ghost of Jupiter, Jupiter Nebula), *10h 24.8m, –18°38'*

NGC 3301 galaxy in Leo (= NGC 3760), *10h 36.9m, +21°53'*

NGC 3351 galaxy in Leo (M95, pair with M96), *10h 44.0m, +11°42'*

NGC 3368 galaxy in Leo (M96, pair with M95), *10h 46.8m, +11°49'*

NGC 3372 bright nebula in Carina (eta (η) Carinae Nebula), *10h 43.8m, –59°52'*

NGC 3379 galaxy in Leo (M105), *10h 47.8m, +12°35'*

NGC 3556 galaxy in Ursa Major (M108), *11h 11.5m, +55°40'*

NGC 3561C galaxy in Ursa Major (Ambartsumian's Knot), *11h 11.6m, +28°43'*

NGC 3587 planetary nebula in Ursa Major (M97, Owl Nebula), *11h 14.8m, +55°01'*

NGC 3623 galaxy in Leo (M65, pair with M66), *11h 18.9m, +13°05'*

NGC 3627 galaxy in Leo (M66, pair with M65), *11h 20.3m, +12°59'*

NGC 3628 galaxy in Leo (Leo Triplet, with M65 and M66), *11h 20.3m, +13°36'*

NGC 3745, 3746, 3748, 3750, 3751, 3753, and 3754 galaxies in Leo (Copeland's Septet), *11h 37.7m, +22°01'; 11h 37.7m, +22°00'; 11h 37.8m, +22°02'; 11h 37.9m, +21°58'; 11h 37.9m, +21°56'; 11h 37.9m, +21°59'; and 11h 37.9m, +21°59'*

NGC 3760 galaxy in Leo (= NGC 3301), *10h 36.9m, +21°53'*

NGC 3918 planetary nebula in Centaurus (Blue Nebula), *11h 50.3m, –57°11'*

NGC 3928 galaxy in Ursa Major (Miniature Spiral Galaxy), *11h 51.8m, +48°41'*

NGC 3992 galaxy in Ursa Major (M109), *11h 57.6m, +53°23'*

NGC 4038 and 4039 galaxies in Corvus (Antennae, Ring-Tail Galaxy, Snorter), *12h 01.9m, –18°52'* and *12h 01.9m, –18°53'*

NGC 4169, 4173, 4174, and 4175 galaxies in Coma Berenices (Box), *12h 12.2m, +29°10'; 12h 12.3m, +29°11'; 12h 12.4m, +29°08';* and *12h 12.5m, +29°09'*

NGC 4192 galaxy in Coma Berenices (M98), *12h 13.8m, +14° 54'*

NGC 4254 galaxy in Coma Berenices (M99, Pinwheel Galaxy),
12h 18.8m, +14° 25'

NGC 4258 galaxy in Canes Venatici (M106), *12h 19.0m, +47° 18'*

NGC 4303 galaxy in Virgo (M61), *12h 21.9m, +4° 28'*

NGC 4321 galaxy in Coma Berenices (M100), *12h 22.9m, +15° 49'*

NGC 4374 galaxy in Virgo (M84), *12h 25.1m, +12° 53'*

NGC 4382 galaxy in Coma Berenices (M85), *12h 25.4m, +18° 11'*

NGC 4406 galaxy in Virgo (M86), *12h 26.2m, +12° 57'*

NGC 4435 and 4438 galaxies in Virgo (Eyes), *12h 27.7m, +13° 05'*
and *12h 27.8m, +13° 01'*

NGC 4472 galaxy in Virgo (M49), *12h 29.8m, +8° 00'*

NGC 4486 galaxy in Virgo (M87, Virgo A Galaxy), *12h 30.8m,
+12° 24'*

NGC 4501 galaxy in Coma Berenices (M88), *12h 32.0m, +14° 25'*

NGC 4517A galaxy in Virgo (Reinmuth 80), *12h 32.5m, +0° 23'*

NGC 4548 galaxy in Coma Berenices (possibly M91), *12h 35.4m,
+14° 30'*

NGC 4552 galaxy in Virgo (M89), *12h 35.7m, +12° 33'*

NGC 4567 and 4568 galaxies in Virgo (Siamese Twins), *12h 36.5m,
+11° 15'* and *12h 36.6m, +11° 14'*

NGC 4569 galaxy in Virgo (M90), *12h 36.8m, +13° 10'*

NGC 4579 galaxy in Virgo (M58), *12h 37.7m, +11° 49'*

NGC 4590 globular cluster in Hydra (M68), *12h 39.5m, –26° 45'*

NGC 4594 galaxy in Virgo (Darklane Galaxy, M104, Sombrero
Galaxy), *12h 40.0m, –11° 37'*

NGC 4621 galaxy in Virgo (M59), *12h 42.0m, +11° 39'*

NGC 4649 galaxy in Virgo (M60), *12h 43.7m, +11° 33'*

NGC 4676 galaxies in Coma Berenices (Mice), *12h 46.1m, +30° 44'*

NGC 4736 galaxy in Canes Venatici (M94), *12h 50.9m, +41° 07'*

NGC 4755 open cluster in Crux (kappa (κ) Crucis Cluster, Jewel
Box), *12h 53.6m, –60° 20'*

NGC 4826 galaxy in Coma Berenices (Blackeye Galaxy, M64),
12h 56.7m, +21° 41'

NGC 5024 globular cluster in Coma Berenices (M53), *13h 12.9m, +18° 10'*

NGC 5055 galaxy in Canes Venatici (M63, Sunflower Galaxy), *13h 15.8m, +42° 02'*

NGC 5128 galaxy in Centaurus (Centaurus A), *13h 25.5m, –43° 01'*

NGC 5139 globular cluster in Centaurus (omega (ω) Centauri Cluster), *13h 26.8m, –47° 29'*

NGC 5194 and 5195 galaxy in Canes Venatici (M51, Lord Rosse's "Nebula," Question Mark Galaxy, Whirlpool Galaxy), *13h 29.9m, +47° 12'* and *13h 30.0m, +47° 16'*

NGC 5216 and 5218 galaxies in Ursa Major (Keenan's System), *13h 32.1m, +62° 42'* and *13h 32.2m, +62° 46'*

NGC 5236 galaxy in Hydra (M83), *13h 37.0m, –29° 52'*

NGC 5272 globular cluster in Canes Venatici (M3), *13h 42.2m, +28° 23'*

NGC 5457 galaxy in Ursa Major (M101, Pinwheel Galaxy), *14h 03.2m, +54° 21'*

NGC 5904 globular cluster in Serpens (M5), *15h 18.6m, +2° 05'*

NGC 6027 galaxies in Serpens (Seyfert's Sextet), *15h 59.2m, +20° 45'*

NGC 6093 globular cluster in Scorpius (M80), *16h 17.0m, –22° 59'*

NGC 6121 globular cluster in Scorpius (M4), *16h 23.6m, –26° 32'*

NGC 6169 open cluster in Norma (mu (μ) Normae Cluster), *16h 34.1m, –44° 03'*

NGC 6171 globular cluster in Ophiuchus (M107), *16h 32.5m, –13° 03'*

NGC 6205 globular cluster in Hercules (Great Cluster in Hercules, Hercules Cluster, M13), *16h 41.7m, +36° 28'*

NGC 6218 globular cluster in Ophicuchus (M12), *16h 47.2m, –1° 57'*

NGC 6254 globular cluster in Ophiuchus (M10), *16h 57.1m, –4° 06'*

NGC 6266 globular cluster in Ophiuchus and Scorpius (M62), *17h 01.2m, –30° 07'*

NGC 6273 globular cluster in Ophiuchus (M19), *17h 02.6m, –26° 16'*

NGC 6302 planetary nebula in Scorpius (Bug Nebula), *17h 13.7m, –37° 06'*

NGC 6309 planetary nebula in Ophiuchus (Box Nebula), *17h 14.1m, –12° 54'*

NGC 6333 globular cluster in Ophiuchus (M9), *17h 19.2m, –18° 31'*

NGC 6341 globular cluster in Hercules (M92), *17h 17.1m, +43° 08'*

NGC 6402 globular cluster in Ophiuchus (M14), *17h 37.6m, –3° 15'*

NGC 6405 open cluster in Scorpius (Butterfly Cluster, M6), *17h 40.1m, –32° 13'*

NGC 6445 planetary nebula in Sagittarius (Little Gem), *17h 49.2m, –20° 01'*

NGC 6475 open cluster in Scorpius (M7), *17h 53.9m, –34° 49'*

NGC 6494 open cluster in Sagittarius (M23), *17h 56.8m, –19° 01'*

NGC 6514 bright nebula in Sagittarius (M20, Trifid Nebula), *18h 02.3m, –23° 02'*

NGC 6523 bright nebula in Sagittarius (Lagoon Nebula, M8), *18h 03.8m, –24° 23'*

NGC 6531 open cluster in Sagittarius (M21), *18h 04.6m, –22° 30'*

NGC 6603 open cluster in Sagittarius (enclosed by M24 Milky Way star cloud), *18h 18.4m, –18° 25'*

NGC 6611 bright nebula in Serpens (Eagle Nebula, M16, Star Queen Nebula), *18h 18.8m, –13° 47'*

NGC 6613 open cluster in Sagittarius (M18), *18h 19.9m, –17° 08'*

NGC 6618 bright nebula in Sagittarius (Checkmark Nebula, Horseshoe Nebula, M17, Omega Nebula, Swan Nebula), *18h 20.8m, –16° 11'*

NGC 6626 globular cluster in Sagittarius (M28), *18h 24.5m, –24° 52'*

NGC 6637 globular cluster in Sagittarius (M69), *18h 31.4m, –32° 21'*

NGC 6656 globular cluster in Sagittarius (M22), *18h 36.4m, –23° 54'*

NGC 6681 globular cluster in Sagittarius (M70), *18h 43.2m, –32° 18'*

NGC 6694 open cluster in Scutum (M26), *18h 45.2m, –9° 24'*

NGC 6705 open cluster in Scutum (M11, Wild Duck Cluster),
18h 51.1m, –6° 16'

NGC 6715 globular cluster in Sagittarius (M54), *18h 55.1m,
–30° 29'*

NGC 6720 planetary nebula in Lyra (M57, Ring Nebula),
18h 53.6m, +33° 02'

NGC 6779 globular cluster in Lyra (M56), *19h 16.6m, +30° 11'*

NGC 6809 globular cluster in Sagittarius (M55), *19h 40.0m,
–30° 58'*

NGC 6822 galaxy in Sagittarius (Barnard's Galaxy), *19h 44.9m,
–14° 48'*

NGC 6826 planetary nebula in Cygnus (Blinking Planetary
Nebula), *19h 44.8m, +50° 31'*

NGC 6838 globular cluster in Sagitta (M71), *19h 53.8m, +18° 47'*

NGC 6853 planetary nebula in Vulpecula (Diabolo Nebula,
Double-Headed Shot, Dumbbell Nebula, M27), *19h 59.6m,
+22° 43'*

NGC 6864 globular cluster in Sagittarius (M75), *20h 06.1m,
–21° 55'*

NGC 6888 bright nebula in Cygnus (Crescent Nebula), *20h 12.0m,
+38° 21'*

NGC 6913 open cluster in Cygnus (M29), *20h 23.9m, +38° 32'*

NGC 6960 bright nebula in Cygnus (Filamentary Nebula,
Lacework Nebula, Veil Nebula with NGC 6992 and 6995),
20h 45.7m, +30° 43'

NGC 6981 globular cluster in Aquarius (M72), *20h 53.5m, –12° 32'*

NGC 6992 bright nebula in Cygnus (Network nebula with NGC
6995, Veil Nebula with NGC 6960 and 6995), *20h 56.4m,
+31° 43'*

NGC 6994 open cluster in Aquarius (M73), *20h 59.0m, –12° 38'*

NGC 6995 bright nebula in Cygnus (Network nebula with NGC
6992, Veil Nebula with NGC 6960 and 6992), *20h 57.1m,
+31° 13'*

NGC 7000 bright nebula in Cygnus (America Nebula, North
America Nebula), *20h 58.8m, +44° 20'*

NGC 7009 planetary nebula in Aquarius (Saturn Nebula), *21h 04.2m, –11°22'*

NGC 7078 globular cluster in Pegasus (M15), *21h 30.0m, +12°10'*

NGC 7088 nebula in Aquarius (Baxendell's Unphotographable Nebula), *21h 33.4m, –0°23'*

NGC 7089 globular cluster in Aquarius (M2), *21h 33.5m, –0°49'*

NGC 7092 open cluster in Cygnus (M39), *21h 32.2m, +48°26'*

NGC 7099 globular cluster in Capricornus (M30), *21h 40.4m, –23°11'*

NGC 7293 planetary nebula in Aquarius (Helical Nebula, Helix Nebula), *22h 29.6m, –20°48'*

NGC 7317 galaxy in Pegasus, *22h 35.9m, +33°57'*

NGC 7318A and 7318B galaxies in Pegasus, *22h 35.9m, +33°58'* and *22h 36.0m, +33°58'*

NGC 7319 galaxy in Pegasus (Stephan's Quartet with NGC 7317, 7318A, and 7318B), *22h 36.1m, +33°59'*

NGC 7320 galaxy in Pegasus (Stephan's Quintet with NGC 7317, 7318A, 7318B, and 7319), *22h 36.1m, +33°57'*

NGC 7552, 7582, 7590, and 7599 galaxies in Grus (Grus Quartet), *23h 16.2m, –42°35'; 23h 18.4m, –42°22'; 23h 18.9m, –42°14'; and 23h 19.3m, –42°15'*

NGC 7635 bright nebula in Cassiopeia (Bubble Nebula), *23h 20.7m, +61°12'*

NGC 7654 open cluster in Cassiopeia (M52), *23h 24.2m, +61°35'*

NGC 7662 planetary nebula in Andromeda (Blue Snowball), *23h 25.9m, +42°33'*

Niebelungen Ring galaxy in Volans (Das Rheingold, Graham's Object), *6h 41.4m, –74°19'*

Nihal (beta (β) Leporis), *5h 28.2m, –20°46'*

Nile Star (alpha (α) Canis Majoris, Dog Star, Sirius), *6h 45.1m, –16°43'*

96 G. Piscium, *0h 48.4m, +5°17'*

Noah's Dove. *See* **Columba**

Nodus Secundus (Altais, delta (δ) Draconis), *19h 12.6m, +67°40'*

Norma (abbreviated **Nor**, Level, Square and Rule):

cluster, open, mu (μ) Normae Cluster (NGC 6169), *16h 34.1m, –44° 03'*

nebula, bright, Ant Nebula, *16h 17.2m, –51° 59'*

nebula, planetary, Shapley 1, *15h 51.7m, –51° 31'*

North America (bright) Nebula in Cygnus (America Nebula, NGC 7000), *20h 58.8m, +44° 20'*

North Star (alpha (α) Ursae Minoris, Cynosura, Polaris, Pole Star), *2h 31.8m, +89° 16'*

Northern Coalsack (dark) Nebula in Cygnus, *20h 40.0m, +42° 00'*

Northern Cross. *See* **Cygnus**

Northern Crown. *See* **Corona Borealis**

nu (ν) Scorpii (Jabbah), *16h 12.0m, –19° 27'*

nu (ν) Ursae Majoris (Alula Borealis), *11h 18.5m, +33° 06'*

Nubecula Major galaxy in Dorado and Mensa (Great Magellanic Cloud, Large Magellanic Cloud, LMC), centered in Dorado at *5h 23.6m, –69° 45'*

Nubecula Minor galaxy in Tucana (Lesser Magellanic Cloud, NGC 292, Small Magellanic Cloud, SMC), *0h 52.7m, –72° 50'*

Nunki (sigma (σ) Sagittarii), *18h 55.3m, –26° 18'*

Nusakan (beta (β) Coronae Borealis), *15h 27.8m, +29° 06'*

Oo

Octans (abbreviated **Oct**, Octant):

> galaxy, Polarissima Australis (NGC 2573), *1h 42.0m, –89°20'*
>
> stars:
>
>> Kurtz's Light Variable, *20h 03.6m, –78°50'*
>>
>> sigma (σ) Octantis (Polaris Australis), *21h 08.8m, –88°57'*

Octant. *See* **Octans**

Okda (alpha (α) Piscium, Alrescha, Alrischa, Kaitain), *2h 02.0m, +2°46'*

Olber's Star in Virgo, *13h 14.1m, –16°33'*

omega (ω) Centauri (globular) Cluster (NGC 5139), *13h 26.8m, –47°29'*

omega (ω) Herculis (Cujam), *16h 25.4m, +14°02'*

Omega (bright) Nebula in Sagittarius (Checkmark Nebula, Horseshoe Nebula, M17, NGC 6618, Swan Nebula), *18h 20.8m, –16°11'*

omicron (o) Canis Majoris (open) Cluster, *6h 54.2m, –24°38'*

omicron (o) Ceti (Mira, Wonderful Star), *2h 19.3m, –2°59'*

omicron[1] (o[1]) Eridani (Beid), *4h 11.9m, –6°50'*

omicron[2] (o[2]) Eridani (Keid), *4h 15.3m, –7°39'*

omicron (o) Ursae Majoris (Muscida), *8h 30.3m, +60°43'*

omicron (o) Velorum (open) Cluster (IC 2391, Velorum), *8h 40.2m, –53°04'*

Ophiuchus (abbreviated **Oph**, Serpent Bearer):

> clusters, globular:
>
>> M9 (NGC 6333), *17h 19.2m, –18°31'*
>>
>> M10 (NGC 6254), *16h 57.1m, –4°06'*
>>
>> M12 (NGC 6218), *16h 47.2m, –1°57'*
>>
>> M14 (NGC 6402), *17h 37.6m, –3°15'*

Ophiuchus (cont.):

    clusters, globular (cont.):

        M19 (NGC 6273), *17h 02.6m, –26°16'*

        M62 (NGC 6266), *17h 01.2m, –30°07'*

        M107 (NGC 6171), *16h 32.5m, –13°03'*

    nebulae, bright, rho (ρ) Ophiuchi Nebula (IC 4604), *16h 25.6m, –23°26'*

    nebulae, dark:

        Barnard's S Nebula (Dark S Nebula, S Nebula, Snake Nebula), *17h 23.5m, –23°38'*

        Pipe Nebula, bowl at *17h 33.0m, –26°00'* and stem at *17h 21.0m, –27°00'*

        rho (ρ) Ophiuchi Dark Cloud, *16h 25.6m, –24°00'*

    nebulae, planetary, Box Nebula (NGC 6309), *17h 14.1m, –12°54'*

    stars:

        alpha (α) Ophiuchi (Rasalague, Rasalhague), *17h 34.9m, +12°34'*

        Barnard's Star (Runaway Star), *17h 57.8m, +4°42'*

        beta (β) Ophiuchi (Cebalrai, Cheleb), *17h 43.5m, +4°34'*

        delta (δ) Ophiuchi (Yed Prior), *16h 14.3m, –3°42'*

        epsilon (ε) Ophiuchi (Yed Posterior), *16h 18.3m, –4°42'*

        eta (η) Ophiuchi (Sabik), *17h 10.4m, –15°44'*

        Hind's New Star, *16h 59.5m, –12°54'*

        Iron Star, *17h 43.9m, –6°16'*

        Kepler's Star, *17h 30.6m, –21°29'*

        lambda (λ) Ophiuchi (Marfak, Marfik), *16h 30.9m, +1°59'*

Orion (abbreviated **Ori**, Giant, Hunter, Warrior):

    clusters, open:

        lambda (λ) Orionis Cluster, *5h 35.1m, +9°56'*

        Trapezium (theta[1] (θ[1]) Orionis' four brightest components), *5h 35.4m, –5°23'*

Orion (cont.):

nebulae, bright:

Barnard's Loop (Orion Loop), centered at *5h 35.0m, –3°00'*

Great Nebula in Orion (M42, NGC 1976, Orion Nebula),
5h 35.4m, –5°27'

Hammerhead Nebula, *5h 33.7m, –5°40'*

Lower's Nebula, *6h 08.9m, +15°49'*

M42 (Great Nebula in Orion, NGC 1976, Orion Nebula),
5h 35.4m, –5°27'

M43 detached portion of M42 (NGC 1982), *5h 35.6m,
–5°16'*

M78 (NGC 2068), *5h 46.7m, +0°03'*

Orion's Cloak, *5h 31.4m, –2°41'*

nebulae, dark:

Fish's Mouth, *5h 35.4m, –5°23'*

Horsehead Nebula (Dark Bay Nebula), *5h 41.0m, –2°24'*

stars:

alpha (α) Orionis (Betelgeuse, Martial Star, Mirzam),
5h 55.2m, +7°24'

Becklin's Star, *5h 35.3m, –5°23'*

beta (β) Orionis (Algebar, Elgebar, Rigel), *5h 14.5m, –8°12'*

Chanal's Variable Star, *5h 34.8m, –5°34'*

Cohen-Schwartz Star, *5h 36.4m, –6°46'*

delta (δ) Orionis (Mintaka), *5h 32.0m, –0°18'*

delta, epsilon, and zeta (δ, ε, and ζ) Orionis (Belt of Orion),
5h, 32.0m, –0°18'; 5h 36.2m, –1°12'; and
5h 40.8m, –1°57'

epsilon (ε) Orionis (Alnilam, Anilam), *5h 36.2m, –1°12'*

gamma (γ) Orionis (Amazon Star, Bellatrix), *5h 25.1m,
+6°21'*

kappa (κ) Orionis (Saiph), *5h 47.8m, –9°40'*

lambda (λ) Orionis (Meissa), *5h 35.1m, +9°56'*

Orion (cont.):

stars (cont.):

Trapezium (theta[1] (θ^1) Orionis' four bright components), 5h 35.4m, –5°23'

Wachmann's Flare Star, 5h 33.8m, +1°57'

zeta (ζ) Orionis (Alnitak), 5h 40.8m, –1°57'

Orion Loop bright nebula in Orion (Barnard's Loop), centered at 5h 35.0m, –3°00'

Orion (bright) Nebula (Great Nebula in Orion, M42, NGC 1976), 5h 35.4m, –5°27'

Orion's Cloak bright nebula, 5h 31.4m, –2°41'

Osawa's Star in Cassiopeia, 23h 32.8m, +57°54'

Owl (open) Cluster in Cassiopeia (NGC 457), 1h 19.1m, +58°20'

Owl (planetary) Nebula in Ursa Major (M97, NGC 3587), 11h 14.8m, +55°01'

Pp

p Carinae, *10h 32.0m, –61° 41'*

P Cygni (34 Cygni), *20h 17.8m, +38° 02'*

p Eridani, *1h 39.8m, –56° 12'*

Painter. *See* **Pictor**

Pair of Compasses. *See* **Circinus**

Palilicium (Aldebaran, alpha (α) Tauri, Cor Tauri), *4h 35.9m, +16° 31'*

Papillon galaxy in Ursa Major (IC 708), *11h 33.9m, +49° 03'*

Parrot's Head (dark) Nebula in Sagittarius, *18h 04.3m, –32° 30'*

Pavo (abbreviated **Pav**, Peacock):

    galaxy, McLeish's Object, *20h 09.7m, –66° 13'*

    star, alpha (α) Pavonis (Peacock Star), *20h 25.6m, –56° 44'*

Pazmino's (open) Cluster in Camelopardalis, *3h 16.3m, +60° 02'*

Peacock. *See* **Pavo**

Peacock Star (alpha (α) Pavonis), *20h 25.6m, –56° 44'*

Pearce's Star (AO Cassiopeiae), *0h 17.7m, +51° 26'*

Pegasus (abbreviated **Peg**, Winged Horse):

    cluster, globular, M15 (NGC 7078), *21h 30.0m, +12° 10'*

    galaxies:

        Barbon's Galaxy, *23h 37.7m, +30° 08'*

        Pegasus Dwarf Galaxy, *23h 28.6m, +14° 45'*

        Pegasus I Cluster, *23h 22.0m, +9° 02'*

        Pegasus II Cluster, *23h 10.0m, +7° 36'*

        Stephan's Quartet (NGC 7317, 7318A, 7318B, and 7319), *22h 35.9m, +33° 57'; 22h 35.9m, +33° 58'; 22h 36.0m, +33° 58';* and *22h 36.1m, +33° 59'*

        Stephan's Quintet (NGC 7317, 7318A, 7318B, 7319, and 7320), *22h 35.9m, +33° 57'; 22h 35.9m, +33° 58'; 22h 36.0m, +33° 58'; 22h 36.1m, +33° 59';* and *22h 36.1m, +33° 57'*

Pegasus (cont.):

    nebula, bright, Moving Nebula, *0h 06.4m, +20° 12'*

    stars:

        alpha (α) Pegasi (Marchab, Markab), *23h 04.8m, +15° 12'*

        alpha, beta, & gamma (α, β, & γ) Pegasi and alpha (α) Andromedae (Great Square of Pegasus), *23h 04.8m, +15° 12'; 23h 03.8m, +28° 05'; & 0h 13.2m, +15° 11'; and 0h 08.4m, +29° 05'*

        beta (β) Pegasi (Scheat), *23h 03.8m, +28° 05'*

        epsilon (ε) Pegasi (Enif, Fumalfaras), *21h 44.2m, +9° 52'*

        eta (η) Pegasi (Matar, Sadalmatar), *22h 43.0m, +30° 13'*

        gamma (γ) Pegasi (Algenib), *0h 13.2m, +15° 11'*

        lambda and mu (λ and μ) Pegasi (Sadalbari), *22h 46.5m, +23° 34' and 22h 50.0m, +24° 36'*

        theta (θ) Pegasi (Baham, Biham), *22h 10.2m, +6° 12'*

        zeta (ζ) Pegasi (Homam), *22h 41.5m, +10° 50'*

Pegasus Dwarf Galaxy, *23h 28.6m, +14° 45'*

Pegasus I Cluster of galaxies, *23h 22.0m, +9° 02'*

Pegasus II Cluster of galaxies, *23h 10.0m, +7° 36'*

Pelican (bright) Nebula in Cygnus (IC 5067 and 5070), *20h 47.8m, +44° 22' and 20h 50.8m, +44° 21'*

Pendulum Clock, Clock (**Horologium**, abbreviated **Hor**)

Perseus (abbreviated **Per**, Champion):

    clusters, open:

        alpha (α) Persei Cluster, *3h 22.0m, +49° 00'*

        Double Cluster (NGC 869 and 884, Sword-Hand of Perseus), *2h 19.0m, +57° 09' and 2h 22.4m, +57° 07'*

        M34 (NGC 1039), *2h 42.0m, +42° 47'*

    galaxy, Perseus A (NGC 1275), *3h 19.8m, +41° 31'*

    nebulae, bright:

        California Nebula (NGC 1499), *4h 00.7m, +36° 37'*

        Chu's Object, *3h 56.8m, +51° 26'*

Perseus (cont.):

nebulae, planetary, Butterfly Nebula (Cork Nebula, Little Dumbbell Nebula, M76, NGC 650 and 651), *1h 42.3m, +51°34'* and *1h 42.3m, +51°35'*

stars:

alpha (α) Persei (Alchemb, Algenib, Marfak, Marfik, Mirfak), *3h 24.3m, +49°52'*

b Persei, *4h 18.2m, +50°18'*

beta (β) Persei (Algol, Demon Star, Medusa Star), *3h 08.2m, +40°57'*

Bidelman's Peculiar Star, *4h 48.9m, +43°17'*

xi (ξ) Persei (Menkib), *3h 59.0m, +35°47'*

zeta (ζ) Persei (Atik), *3h 54.1m, +31°53'*

Perseus A galaxy (NGC 1275), *3h 19.8m, +41°31'*

Phact (alpha (α) Columbae, Phaet), *5h 39.6m, –34°04'*

Phad (gamma (γ) Ursae Majoris, Phaed, Phecda), *11h 53.8m, +53°42'*

Phaet (alpha (α) Columbae, Phact), *5h 39.6m, –34°04'*

Phe. *See* **Phoenix**

Phecda (gamma (γ) Ursae Majoris, Phad, Phaed), *11h 53.8m, +53°42'*

Pherkad (gamma (γ) Ursae Minoris), *15h 20.7m, +71°50'*

Phoenix (abbreviated **Phe**):

galaxy, Exclamation Mark Galaxy, *0h 39.3m, –43°06'*

star, alpha (α) Phoenicis (Ankaa), *0h 26.3m, –42°18'*

Phurud (Furud, zeta (ζ) Canis Majoris), *6h 20.3m, –30°04'*

pi^1 (π^1) Cygni (Azelfafage), *21h 42.1m, +51°11'*

Piazzi's Flying Star (Flying Star, 61 Cygni), *21h 06.9m, +38°45'*

Pic. *See* **Pictor**

Pickering's Triangular Wisp bright nebula in Cygnus (part of Veil Nebula), *20h 48.5m, +31°09'*

Pictor (abbreviated **Pic**, Easel, Painter):

star, Kapteyn's Star, *5h 10.6m, –44°52'*

Pinwheel Galaxy:

> M33 in Triangulum (NGC 598, Triangulum Galaxy), *1h 33.9m, +30° 39'*

> M99 in Coma Berenices (NGC 4254), *12h 18.8m, +14° 25'*

> M101 in Ursa Major (NGC 5457), *14h 03.2m, +54° 21'*

Pipe (dark) Nebula in Ophiuchus, bowl at *17h 33.0m, –26° 00'* and stem at *17h 21.0m, –27° 00'*

Pisces (abbreviated **Psc**, Circlet, Fishes):

> galaxy, M74 (NGC 628), *1h 36.7m, +15° 47'*

> stars:

>> alpha (α) Piscium (Alrescha, Alrischa, Kaitain, Okda), *2h 02.0m, +2° 46'*

>> 96 G. Piscium, *0h 48.4m, +5° 17'*

>> Van Maanen's Star, *0h 49.2m, +5° 23'*

Piscis Austrinus (abbreviated **PsA**, Southern Fish):

> stars:

>> alpha (α) Piscis Austrini (First Frog, Fomalhaut), *22h 57.6m, –29° 37'*

>> Lacaille 9352, *23h 05.9m, –35° 51'*

Piscis Volantis. *See* **Volans**

Plaskett's Star in Monoceros, *6h 37.5m, +6° 08'*

Pleiades open cluster in Taurus (M45, Seven Sisters), *3h 47.0m, +24° 07'*

Pleione (28 Tauri), *3h 49.2m, +24° 08'*

Pleione (bright) Nebula in Taurus, *3h 49.2m, +24° 08'*

Plough. *See* **Ursa Major**

Pointer Stars (alpha and beta (α and β) Ursae Majoris), *11h 03.7m, +61° 45'* and *11h 01.8m, +56° 23'*

Polaris (alpha (α) Ursae Minoris, Cynosura, North Star, Pole Star), *2h 31.8m, +89° 16'*

Polaris Australis (sigma (σ) Octantis), *21h 08.8m, –88° 57'*

Polarissima Australis galaxy in Octans (NGC 2573), *1h 42.0m, –89° 20'*

Pp

Polarissima Borealis galaxy in Ursa Minor (NGC 3172), *11h 50.0m, +89° 07'*

Pole Star (alpha (α) Ursae Minoris, Cynosura, North Star, Polaris), *2h 31.8m, +89° 16'*

Pollux (beta (β) Geminorum), *7h 45.3m, +28° 02'*

Pollux and Castor. *See* **Gemini**

Poop of Argo Navis. *See* **Puppis**

Popper's Star in Centaurus, *14h 15.0m, –46° 17'*

Porrima (Antevorta, gamma (γ) Virginis), *12h 41.7m, –1° 27'*

Praesepe (open) Cluster in Cancer (Beehive Cluster, M44, Manger, NGC 2632), *8h 40.1m, +19° 59'*

Procyon (alpha (α) Canis Minoris, Little Dog Star), *7h 39.3m, +5° 14'*

Propus (eta (η) Geminorum), *6h 14.9m, +22° 30'*

Proxima Centauri (Inne's Star), *14h 29.7m, –62° 41'*

Przybylski's Star in Centaurus, *11h 37.6m, –46° 43'*

PsA. *See* **Piscis Austrinus**

Psc. *See* **Pisces**

psi (ψ) Draconis (Dsiban), *17h 41.9m, +72° 09'*

Pulcherrima (epsilon (ε) Bootis, Izar, Mizar), *14h 45.0m, +27° 04'*

Pump, Air Pump (**Antlia**, abbreviated **Ant**)

Pup (Sirius B), *6h 45.2m, –16° 43'*

Puppis (abbreviated **Pup**, Poop or Stern of Argo Navis):

    clusters, open:

        M46 (NGC 2437), *7h 41.8m, –14° 49'*

        M47 (NGC 2422), *7h 36.6m, –14° 30'*

        M93 (NGC 2447), *7h 44.6m, –23° 52'*

    stars:

        L^1 Puppis, *7h 13.2m, –45° 11'*

        L^2 Puppis, *7h 13.5m, –44° 38'*

        zeta (ζ) Puppis (Naos), *8h 03.6m, –40° 00'*

Pyxis (abbreviated **Pyx**, Compass of Argo Navis, Mariner's Compass) q Carinae, *10h 17.1m, –61°20'*

Question Mark Galaxy in Canes Venatici (Lord Rosse's "Nebula," M51, NGC 5194 and 5195, Whirlpool Galaxy), *13h 29.9m, +47°12'* and *13h 30.0m, +47°16'*

Rr

R Leporis (Crimson Star, Hind's Crimson Star), *4h 59.6m, –14° 48'*

Ram. *See* **Aries**

Rasaben (Eltanin, Etamin, Ettanin, gamma (γ) Draconis,
Rastaban), *17h 56.6m, +51° 29'*

Rasalague (Rasalhague, alpha (α) Ophiuchi), *17h 34.9m, +12° 34'*

Rasalas (Alshemali, mu (μ) Leonis), *9h 52.8m, +26° 00'*

Rasalgethi (alpha¹ (α¹) Herculis), *17h 14.6m, +14° 23'*

Rasalhague (alpha (α) Ophiuchi, Rasalague), *17h 34.9m, +12° 34'*

Rasalmuthallath (alpha (α) Trianguli, Caput Trianguli), *1h 53.1m,
+29° 35'*

Rastaban:

> beta (β) Draconis (Alwaid), *17h 30.4m, +52° 18'*

> gamma (γ) Draconis (Eltanin, Etamin, Ettanin, Rasaben),
> *17h 56.6m, +51° 29'*

Raven. *See* **Corvus**

Red Bird (Alfard, alpha (α) Hydrae, Alphard, Cor Hydrae),
9h 27.6m, –8° 40'

Red Rectangle planetary nebula in Monoceros, *6h 20.0m, –10° 39'*

Regor (gamma² (γ²) Velorum, Spectral Gem of the Southern Skies,
Suhail), *8h 09.5m, –47° 20'*

Regulus (alpha (α) Leonis, Cor Leonis), *10h 08.4m, +11° 58'*

Regulus Dwarf Galaxy in Leo (Leo I), *10h 08.4m, +12° 18'*

Reinmuth 80 galaxy in Virgo (NGC 4517A), *12h 32.5m, +0° 23'*

Reticulum (abbreviated **Ret**, Net, Rhomboidal Net)

rho (ρ) Ophiuchi Dark Cloud (nebula), *16h 25.6m, –24° 00'*

rho (ρ) Ophiuchi (bright) Nebula (IC 4604), *16h 25.6m, –23° 26'*

Rhomboidal Net, Net (**Reticulum**, abbreviated **Ret**)

Rigel (Algebar, beta (β) Orionis, Elgebar), *5h 14.5m, –8° 12'*

Rigil Kentaurus (alpha[1] (α^1) Centauri, Toliman), *14h 39.6m,
        –60° 50'*

Ring (planetary) Nebula in Lyra (M57, NGC 6720), *18h 53.6m,
        +33° 02'*

Ring-Tail Galaxy in Corvus (Antennae, NGC 4038 and 4039,
        Snorter), *12h 01.9m, –18° 52'* and *12h 01.9m, –18° 53'*

River. *See* **Eridanus**

Roberts-Altizer variable star in Draco, *10h 15.6m, +73° 26'*

Rosette (bright) Nebula in Monoceros (NGC 2237, 2238, 2239, and
        2246), *6h 30.3m, +5° 03'; 6h 30.6m, +5° 01'; 6h 31.0m,
        +4° 57'; and 6h 32.4m, +5° 07'*

Rosino-Zwicky variable star in Coma Berenices, *12h 32.4m,
        +14° 19'*

Rotanev (beta (β) Delphini), *20h 37.5m, +14° 36'*

Ruchbah (delta (δ) Cassiopeiae), *1h 25.8m, +60° 14'*

Rukbat (alpha (α) Sagittarii), *19h 23.9m, –40° 37'*

Runaway Star:

    Barnard's Star in Ophiuchus, *17h 57.8m, +4° 42'*

    Groombridge 1830 in Ursa Major, *11h 53.0m, +37° 43'*

Running Chicken (bright) Nebula in Centaurus (IC 2944),
        11h 36.6m, –63° 02'

Ss

S (dark) Nebula in Ophiuchus (Barnard's S Nebula, Dark S
    Nebula, Snake Nebula), *17h 23.5m, –23°38'*

Sabik (eta (η) Ophiuchi), *17h 10.4m, –15°44'*

Sadachbia (gamma (γ) Aquarii), *22h 21.7m, –1°23'*

Sadalbari (lambda and mu (λ and μ) Pegasi), *22h 46.5m, +23°34'*
    and *22h 50.0m, +24°36'*

Sadalmatar (eta (η) Pegasi, Matar), *22h 43.0m, +30°13'*

Sadalmelik (alpha (α) Aquarii), *22h 05.8m, –0°19'*

Sadalsuud (beta (β) Aquarii), *21h 31.6m, –5°34'*

Sadatoni (zeta (ζ) Aurigae), *5h 02.5m, +41°05'*

Sadr (gamma (γ) Cygni), *20h 22.2m, +40°15'*

Sagitta (abbreviated **Sge**, Arrow):

    cluster, globular, M71 (NGC 6838), *19h 53.8m, +18°47'*

    star, Merrill's Star, *19h 11.5m, +16°52'*

Sagittarius (abbreviated **Sgr**, Archer, Milk Dipper, Teapot):

    clusters, globular:

        M22 (NGC 6656), *18h 36.4m, –23°54'*

        M28 (NGC 6626), *18h 24.5m, –24°52'*

        M54 (NGC 6715), *18h 55.1m, –30°29'*

        M55 (NGC 6809), *19h 40.0m, –30°58'*

        M69 (NGC 6637), *18h 31.4m, –32°21'*

        M70 (NGC 6681), *18h 43.2m, –32°18'*

        M75 (NGC 6864), *20h 06.1m, –21°55'*

    clusters, open:

        M18 (NGC 6613), *18h 19.9m, –17°08'*

        M21 (NGC 6531), *18h 04.6m, –22°30'*

        M23 (NGC 6494), *17h 56.8m, –19°01'*

        M25 (IC 4725), *18h 31.6m, –19°15'*

        NGC 6603 (enclosed by M24 Milky Way star cloud),
            18h 18.4m, –18°25'

Sagittarius (cont.):

    galaxies:

        Barnard's Galaxy (NGC 6822), *19h 44.9m, –14° 48'*

        Milky Way centered at *17h 45.6m, –28° 56'*

        Sagittarius Dwarf Galaxy, *19h 30.0m, –17° 41'*

    nebulae, bright:

        Hourglass Nebula, *18h 03.8m, –24° 23'*

        M8 (Lagoon Nebula, NGC 6523), *18h 03.8m, –24° 23'*

        M17 (Checkmark Nebula, Horseshoe Nebula, NGC 6618, Omega Nebula, Swan Nebula), *18h 20.8m, –16° 11'*

        M20 (NGC 6514, Trifid Nebula), *18h 02.3m, –23° 02'*

    nebulae, dark:

        Black Hole, *18h 15.5m, –18° 11'*

        Dragon, *18h 04.8m, –24° 30'*

        Parrot's Head Nebula, *18h 04.3m, –32° 30'*

    nebulae, planetary, Little Gem (NGC 6445), *17h 49.2m, 20° 01'*

    stars:

        alpha (α) Sagittarii (Rukbat), *19h 23.9m, –40° 37'*

        beta[1] (β[1]) Sagittarii (Arkab), *19h 22.6m, –44° 28'*

        Bond's Flare Star, *23h 31.7m, –2° 45'*

        delta (δ) Sagittarii (Media, Kaus Meridionalis), *18h 21.0m, –29° 50'*

        epsilon (ε) Sagittarii (Kaus Australis), *18h 24.2m, –34° 23'*

        gamma (γ) Sagittarii (Alnasl), *18h 05.8m, –30° 25'*

        lambda (λ) Sagittarii (Kaus Borealis), *18h 28.0m, –25° 25'*

        M24 star cloud (Delle Caustiche, Small Sagittarius Cloud), *18h 16.9m, –18° 29'*

        sigma (σ) Sagittarii (Nunki), *18h 55.3m, –26° 18'*

        zeta (ζ) Sagittarii (Ascella), *19h 02.6m, –29° 53'*

Sagittarius Dwarf Galaxy, *19h 30.0m, –17° 41'*

Saidak (Alcor, 80 Ursae Majoris, g Ursae Majoris), *13h 25.2m, +54° 59'*

Sails of Argo Navis. *See* **Vela**

Saiph (kappa (κ) Orionis), *5h 47.8m, –9° 40'*

Sanduleak-Pesch star in Hercules, *17h 05.5m, +48° 03'*

Sanduleak-Stephenson star in Aquila (SS 433), *19h 11.8m, +4° 59'*

Sanduleak's Star in Mensa, *5h 45.3m, –71° 16'*

Sargas (theta (θ) Scorpii), *17h 37.3m, –43° 00'*

Saturn (planetary) Nebula in Aquarius (NGC 7009), *21h 04.2m, –11° 22'*

Scales. *See* **Libra**

Schaeberle's Flaming Star bright nebula in Auriga (Flaming Star Nebula, IC 405), *5h 16.2m, +34° 16'*

Scheat:

    beta (β) Pegasi, *23h 03.8m, +28° 05'*

    delta (δ) Aquarii (Skat), *22h 54.6m, –15° 49'*

Schedar (alpha (α) Cassiopeiae, Schedir, Shedar), *0h 40.5m, +56° 32'*

Scheiner's Star in Leo, *9h 37.2m, +15° 14'*

Schemali (Deneb Kaitos, iota (ι) Ceti), *0h 19.4m, –8° 49'*

Schweizer-Middleditch Star in Centaurus, *15h 02.8m, –41° 59'*

Scl. *See* **Sculptor**

Scorpius (abbreviated **Sco**, Scorpion):

    clusters, globular:

        M4 (NGC 6121), *16h 23.6m, –26° 32'*

        M62 (NGC 6266), *17h 01.2m,–30° 07'*

        M80 (NGC 6093), *16h 17.0m, –22° 59'*

    clusters, open:

        M6 (Butterfly Cluster, NGC 6405), *17h 40.1m, –32° 13'*

        M7 (NGC 6475), *17h 53.9m, –34° 49'*

    nebula, bright, Antares Nebula (Cloud Nebula), *16h 29.2m, –26° 27'*

    nebula, planetary, Bug Nebula (NGC 6302), *17h 13.7m, –37° 06'*

Scorpius (cont.):

    stars:

        alpha (α) Scorpii (Antares, Cor Scorpii), *16h 29.4m, –26° 26'*

        beta[1] (β[1]) Scorpii (Acrab, Akrab, Graffias), *16h 05.4m, –19° 48'*

        delta (δ) Scorpii (Dschubba), *16h 00.3m, –22° 37'*

        G Scorpii, *17h 49.9m, –37° 03'*

        lambda (λ) Scorpii (Shaula), *17h 33.6m, –37° 06'*

        nu (ν) Scorpii (Jabbah), *16h 12.0m, –19° 27'*

        sigma (σ) Scorpii (Alniyat), *16h 21.2m, –25° 36'*

        theta (θ) Scorpii (Sargas), *17h 37.3m, –43° 00'*

        upsilon (υ) Scorpii (Lesath, Lesuth), *17h 30.8m, –37° 18'*

Sct. *See* **Scutum**

Sculptor (abbreviated **Scl**):

    cluster, open, zeta (ζ) Sculptoris Cluster, *0h 04.3m, –29° 56'*

    galaxies:

        Cartwheel Galaxy, *0h 37.4m, –33° 45'*

        Sculptor Dwarf Galaxy, *0h 59.9m, –33° 42'*

        Sculptor Dwarf Irregular Galaxy (SDIG), *0h 08.1m, –34° 34'*

        Sculptor Galaxy (NGC 253), *0h 47.6m, –25° 17'*

        Sculptor Group, *0h 38.0m, –30° 58'*

    star, Eggen's Nearby Star, *1h 32.3m, –30° 41'*

Scutum (abbreviated **Sct**, Shield, Sobieski's Shield):

    clusters, open:

        M11 (NGC 6705, Wild Duck Cluster), *18h 51.1m, –6° 16'*

        M26 (NGC 6694), *18h 45.2m, –9° 24'*

    stars:

        Branchett's Object, *18h 46.9m, –4° 57'*

        Gem of the Milky Way (Scutum Star Cloud), *18h 40.0m, –7° 00'*

SDIG (Sculptor Dwarf Irregular Galaxy), *0h 08.1m, –34° 34'*

Sea Monster. *See* **Cetus**

Sea Serpent. *See* **Hydra**

Seagoat. *See* **Capricornus**

Seagull (bright) Nebula in Dorado (NGC 2032, 2033, 2034, and 2035), *5h 35.3m, –67° 34'; 5h 34.5m, –69° 44'; 5h 35.7m, –66° 56'; and 5h 35.3m, –67° 34'*

Seashell Galaxy in Centaurus, *13h 47.4m, –30° 25'*

Second Frog (beta (β) Ceti, Deneb Kaitos, Difda, Diphda), *0h 43.6m, –17° 59'*

Seginus (gamma (γ) Bootis), *14h 32.1m, +38° 19'*

Serpens (abbreviated **Ser**, Serpent, Serpens Caput (the head) and Serpens Cauda (the tail) together):

cluster, globular, M5 (NGC 5904), *15h 18.6m, +2° 05'*

galaxies:

Serpens Dwarf Galaxy, *15h 16.1m, +0° 08'*

Seyfert's Sextet (NGC 6027), *15h 59.2m, +20° 45'*

nebula, bright, M16 (Eagle Nebula, NGC 6611, Star Queen Nebula), *18h 18.8m, –13° 47'*

stars:

alpha (α) Serpentis (Cor Serpentis, Unukalhai), *15h 44.3m, +6° 26'*

d Serpentis (59 Serpentis), *18h 27.2m, +0° 12'*

Stepanian's Star, *15h 38.0m, +18° 52'*

theta (θ) Serpentis (Alya), *18h 56.2m, +4° 12'*

Tweedledee and Tweedledum double-double star, *18h 45.5m, +5° 30'*

Serpens Dwarf Galaxy, *15h 16.1m, +0° 08'*

Serpent. *See* **Serpens**

Serpent Bearer. *See* **Ophiuchus**

Serpha (beta (β) Leonis, Dafira, Deneb, Denebola), *11h 49.1m, +14° 34'*

Seven Sisters open cluster in Taurus (M45, Pleiades), *3h 47.0m, +24° 07'*

Sextans (abbreviated **Sex**, Sextant):
>> galaxies:
>>> Sextans A, *10h 11.1m, –4°43'*
>>> Sextans B, *10h 00.0m, +5°20'*
>>> Sextans C, *10h 05.6m, +0°04'*
>>> Spindle "Nebula" (NGC 3115), *10h 05.2m, –7°43'*

Sextant. *See* **Sextans**

Seyfert's Sextet of galaxies in Serpens (NGC 6027), *15h 59.2m, +20°45'*

Sge. *See* **Sagitta**

Sgr. *See* **Sagittarius**

Shakhbazian 1 group of galaxies in Ursa Major, *10h 54.8m, +40°28'*

Shapley 1 planetary nebula in Norma, *15h 51.7m, –51°31'*

Shaula (lambda (λ) Scorpii), *17h 33.6m, –37°06'*

Shedar (alpha (α) Cassiopeiae, Schedar, Schedir), *0h 40.5m, +56°32'*

Sheliak (beta (β) Lyrae, Shelyak), *18h 50.1m, +33°22'*

Sheratan (beta (β) Arietis), *1h 54.6m, +20°48'*

Shield. *See* **Scutum**

Ship Argo. *See* **Carina**; **Puppis**; **Pyxis**; **Vela**

Siamese Twins galaxies in Virgo (NGC 4567 and 4568), *12h 36.5m, +11°15'* and *12h 36.6m, +11°14'*

Sickle. *See* **Leo**

Sidus Ludovicianum star in Ursa Major, *13h 25.0m, +54°53'*

sigma (σ) Octantis (Polaris Australis), *21h 08.8m, –88°57'*

sigma (σ) Sagittarii (Nunki), *18h 55.3m, –26°18'*

sigma (σ) Scorpii (Alniyat), *16h 21.3m, –25°36'*

Sirius (alpha (α) Canis Majoris, Dog Star, Nile Star), *6h 45.1m, –16°43'*

Sirius B (Pup), *6h 45.2m, –16°43'*

Sirrah (Alferats, alpha (α) Andromedae, Alpheratz), *0h 08.4m, +29°05'*

Situla (kappa (κ) Aquarii), *22h 37.8m, –4° 14'*

61 Cygni (Flying Star, Piazzi's Flying Star), *21h 06.9m, +38° 45'*

Skat (delta (δ) Aquarii, Scheat), *22h 54.6m, –15° 49'*

Small Bear. *See* **Ursa Minor**

Small Dog. *See* **Canis Minor**

Small Lion. *See* **Leo Minor**

Small Magellanic Cloud galaxy in Tucana (Lesser Magellanic Cloud, NGC 292, Nubecula Minor, SMC), *0h 52.7m, –72° 50'*

Small Sagittarius (star) Cloud (Delle Caustiche, M24), *18h 16.9m, –18° 29'*

SMC galaxy in Tucana (Lesser Magellanic Cloud, NGC 292, Nubecula Minor, Small Magellanic Cloud), *0h 52.7m, –72° 50'*

Snake (dark) Nebula in Ophiuchus (Barnard's S Nebula, Dark S Nebula, S Nebula), *17h 23.5m, –23° 38'*

Snakes:

    Sea Serpent. *See* **Hydra**

    Serpent. *See* **Serpens**

    Water Monster. *See* **Hydra**

    Water Snake (**Hydrus**, abbreviated **Hyi**)

Snickers galaxy in Gemini, *6h 28.0m, +15° 00'*

Snorter galaxies in Corvus (Antennae, NGC 4038 and 4039, Ring-Tail Galaxy), *12h 01.9m, –18° 52'* and *12h 01.9m, –18° 53'*

Sobieski's Shield. *See* **Scutum**

Sombrero Galaxy in Virgo (Darklane Galaxy, M104, NGC 4594), *12h 40.0m, –11° 37'*

Southern Cross. *See* **Crux**

Southern Crown (**Corona Australis,** abbreviated **CrA**)

Southern Fish. *See* **Piscis Austrinus**

Southern Pleiades open cluster in Carina (IC 2602, theta (θ) Carinae Cluster), *10h 43.2m, –64° 24'*

Southern Triangle. *See* **Triangulum Australe**

Spectral Gem of the Southern Skies (gamma^2 (γ^2) Velorum, Regor, Suhail), *8h 09.5m, –47°20'*

Spica (alpha (α) Virginis, Azimech), *13h 25.2m, –11°10'*

Spider galaxy in Leo Minor, *10h 42.7m, +34°26'*

Spindle "Nebula" galaxy in Sextans (NGC 3115), *10h 05.2m, –7°43'*

Square and Rule. *See* **Norma**

Star Queen (bright) Nebula in Serpens (Eagle Nebula, M16, NGC 6611), *18h 18.8m, –13°47'*

Stepanian's Star in Serpens, *15h 38.0m, +18°52'*

Stephan's Quartet of galaxies in Pegasus (NGC 7317, 7318A, 7318B, and 7319), *22h 35.9m, +33°57'; 22h 35.9m, +33°58'; 22h 36.0m, +33°58'; and 22h 36.1m, +33°59'*

Stephan's Quintet of galaxies in Pegasus (NGC 7317, 7318A, 7318B, 7319, and 7320), *22h 35.9m, +33°57'; 22h 35.9m, +33°58'; 22h 36.0m, +33°58'; 22h 36.1m, +33°59' ; and 22h 36.1m, +33°57'*

Stern of Argo Navis. *See* **Puppis**

Sterope (Asterope, 21 Tauri), *3h 45.9m, +24°33'*

Sterope (bright) Nebula in Taurus, *3h 45.9m, +24°33'*

Struve's Lost (bright) Nebula in Taurus (NGC 1554), *4h 21.8m, +19°32'*

Sualocin (alpha (α) Delphini, Svalocin), *20h 39.6m, +15°55'*

Suhail:

gamma^2 (γ^2) Velorum (Regor, Spectral Gem of the Southern Skies), *8h 09.5m, –47°20'*

lambda (λ) Velorum, *9h 08.0m, –43°26'*

Sulafat (gamma (γ) Lyrae, Sulaphat), *18h 58.9m, +32°41'*

Summer Triangle (alpha (α) Aquilae, alpha (α) Cygni, and alpha (α) Lyrae), *19h 50.8m, +8°52'; 20h 41.4m, +45°17'; and 18h 36.9m, +38°47'*

Sunflower Galaxy in Canes Venatici (M63, NGC 5055), *13h 15.8m, +42°02'*

Svalocin (alpha (α) Delphini, Sualocin), *20h 39.6m, +15°55'*

Swan. *See* **Cygnus**

Swan (bright) Nebula in Sagittarius (Checkmark Nebula, Horseshoe Nebula, M17, NGC 6618, Omega Nebula), *18h 20.8m, –16° 11'*

Sword-Hand of Perseus open cluster (Double Cluster, NGC 869 and 884), *2h 19.0m, +57° 09'* and *2h 22.4m, +57° 07'*

Swordfish. *See* **Dorado**

Syrma (iota (ι) Virginis), *14h 16.0m, –6° 00'*

T Coronae Borealis (Blaze Star), *15h 59.5m, +25°55'*

Table Mountain. *See* **Mensa**

Talita (Dnoces, iota (ι) Ursae Majoris, Talitha), *8h 59.2m, +48°02'*

Tania Australis (mu (μ) Ursae Majoris), *10h 22.3m, +41°30'*

Tania Borealis (lambda (λ) Ursae Majoris), *10h 17.1m, +42°55'*

Tarantula (bright) Nebula in Dorado (Great Looped Nebula, Loop
      Nebula, NGC 2070, 30 Doradus Nebula), *5h 38.7m,*
      –69°06'

Tarazed (gamma (γ) Aquilae), *19h 46.3m, +10°37'*

Taurus (abbreviated **Tau**, Bull):

    clusters, open:

        Hyades (Melotte 25), *4h 26.9m, +15°52'*

        M45 (Pleiades, Seven Sisters), *3h 47.0m, +24°07'*

    galaxy, BW Tauri Galaxy, *4h 33.2m, +5°21'*

    nebulae, bright:

        Alcyone Nebula, *3h 47.5m, +24°06'*

        Atlas Nebula, *3h 49.2m, +24°03'*

        Burnham's Nebula, *4h 22.0m, +19°32'*

        Celaeno Nebula, *3h 44.8m, +24°17'*

        Crab Nebula (M1, NGC 1952), *5h 34.5m, +22°01'*

        Electra Nebula, *3h 44.9m, +24°07'*

        Hind's Variable Nebula (NGC 1555), *4h 22.9m, +19°32'*

        M1 (Crab Nebula, NGC 1952), *5h 34.5m, +22°01'*

        Maia Nebula (NGC 1432), *3h 45.8m, +24°22'*

        Merope Nebula (NGC 1435), *3h 46.1m, +23°47'*

        Pleione Nebula, *3h 49.2m, +24°08'*

        Sterope Nebula, *3h 45.9m, +24°33'*

        Struve's Lost Nebula (NGC 1554), *4h 21.8m, +19°32'*

Taurus (cont.):

    nebulae, bright (cont.):

        Taygeta Nebula, *3h 45.2m, +24° 28'*

        Tempel's Nebula, *3h 46.1m, +23° 47'*

    nebulae, dark:

        Kutner's Cloud, *4h 33.0m, +24° 36'*

        Taurus Dark Cloud, *4h 30.0m, +27° 00'*

    stars:

        alpha (α) Tauri (Aldebaran, Cor Tauri, Palilicium), *4h 35.9m, +16° 31'*

        Alcyone (eta (η) Tauri), *3h 47.5m, +24° 06'*

        Asterope (Sterope, 21 Tauri), *3h 45.9m, +24° 33'*

        Atlas (27 Tauri), *3h 49.2m, +24° 03'*

        beta (β) Tauri (Alnath, Elnath, Nath), *5h 26.3m, +28° 36'*

        Celaeno (16 Tauri), *3h 44.8m, +24° 17'*

        delta[1] (δ^1) Tauri (Hyadum II), *4h 22.9m, +17° 33'*

        Electra (17 Tauri), *3h 44.9m, +24° 07'*

        epsilon (ε) Tauri (Ain), *4h 28.6m, +19° 11'*

        eta (η) Tauri (Alcyone), *3h 47.5m, +24° 06'*

        gamma (γ) Tauri (Hyadum I), *4h 19.8m, +15° 38'*

        Maia (20 Tauri), *3h 45.8m, +24° 22'*

        Merope (23 Tauri), *3h 46.3m, +23° 57'*

        Pleione (28 Tauri), *3h 49.2m, +24° 08'*

        Sterope (Asterope, 21 Tauri), *3h 45.9m, +24° 33'*

        Taygeta (19 Tauri), *3h 45.2m, +24° 28'*

Taurus Dark Cloud dark nebula, *4h 30.0m, +27° 00'*

Taygeta (19 Tauri), *3h 45.2m, +24° 28'*

Taygeta (bright) Nebula in Taurus, *3h 45.2m, +24° 28'*

Teapot. *See* **Sagittarius**

Tegmen (zeta (ζ) Cancri), *8h 12.2m, +17° 39'*

Tejat (mu (μ) Geminorum), *6h 23.0m, +22° 31'*

Telescopium (abbreviated **Tel**, Telescope)

Tempel's (bright) Nebula in Taurus, *3h 46.1m, +23°47'*

theta (θ) Aquarii (Ancha), *22h 16.8m, –7°47'*

theta (θ) Bootis (Asellus), *14h 25.2m, +51°51'*

theta (θ) Carinae open cluster (IC 2602, Southern Pleiades), *10h 43.2m, –64°24'*

theta (θ) Cassiopeiae (Marfak, Marfik), *1h 11.1m, +55°09'*

theta (θ) Centauri (Menkent), *14h 06.7m, –36°22'*

theta[1] (θ[1]) Eridani (Acamar), *2h 58.3m, –40°18'*

theta (θ) Leonis (Chertan, Chort), *11h 14.2m, +15°26'*

theta[1] (θ[1]) Orionis' four bright components (Trapezium), *5h 35.4m, –5°23'*

theta (θ) Pegasi (Baham, Biham), *22h 10.2m, +6°12'*

theta (θ) Scorpii (Sargas), *17h 37.3m, –43°00'*

theta[1] (θ[1]) Serpentis (Alya), *18h 56.2m, +4°12'*

30 Doradus (bright) Nebula (Great Looped Nebula, Loop Nebula, NGC 2070, Tarantula Nebula), *5h 38.7m, –69°06'*

33 G. Librae, *14h 57.5m, –21°25'*

Thuban (alpha (α) Draconis), *14h 04.4m, +64°23'*

Thumbprint (dark) Nebula in Camelopardalis, *12h 42.9m, +78°16'*

Tiger. *See* **Lynx**

Toby Jug (bright) Nebula in Carina (Butterfly Nebula, IC 2220), *7h 56.8m, –59°07'*

Toliman (alpha[1] (α[1]) Centauri, Rigil Kentaurus), *14h 39.6m, –60°50'*

Toucan. *See* **Tucana**

TrA. *See* **Triangulum Australe**

Trapezium open cluster in Orion (theta[1] (θ[1]) Orionis' four bright components), *5h 35.4m, –5°23'*

Tri. *See* **Triangulum**

Triangles. *See* Summer Triangle; **Triangulum; Triangulum Australe**

Triangulum (abbreviated **Tri**, Triangle):

> galaxy, M33 (NGC 598, Pinwheel Galaxy, Triangulum Galaxy), *1h 33.9m, +30° 39'*

> star, alpha (α) Trianguli (Caput Trianguli, Rasalmuthallath), *1h 53.1m, +29° 35'*

Triangulum Australe (abbreviated **TrA**, Southern Triangle):

> star, alpha (α) Trianguli Australis (Atria), *16h 48.7m, –69° 02'*

Triangulum Galaxy (M33, NGC 598, Pinwheel Galaxy), *1h 33.9m, +30° 39'*

Trifid (bright) Nebula in Sagittarius (M20, NGC 6514), *18h 02.3m, –23° 02'*

Tucana (abbreviated **Tuc**, Toucan):

> cluster, globular, 47 Tucanae (NGC 104), *0h 24.1m, –72° 05'*

> galaxy, Small Magellanic Cloud (Lesser Magellanic Cloud, NGC 292, Nubecula Minor, SMC), *0h 52.7m, –72° 50'*

Tweedledee and Tweedledum double-double star in Serpens, *18h 45.5m, +5° 30'*

Twins. *See* **Gemini**

268 G. Ceti, *2h 36.1m, +6° 53'*

Tycho's Star (B Cassiopeiae), *0h 25.3m, +64° 09'*

u Carinae, *10h 53.5m, –58°51'*

u Herculis (68 Herculis), *17h 17.3m, +33°06'*

UMa. *See* **Ursa Major**

UMi. *See* **Ursa Minor**

Unicorn. *See* **Monoceros**

Unukalhai (alpha (α) Serpentis, Cor Serpentis), *15h 44.3m, +6°26'*

upsilon (υ) Scorpii (Lesath, Lesuth), *17h 30.8m, –37°18'*

Ursa Major (abbreviated **UMa**, Big Dipper, Charles's Wain, Greater Bear, Plough):

cluster, open, Ursa Major Moving Cluster, *12h 03.0m, +58°00'*

galaxies:

Ambartsumian's Knot (NGC 3561C), *11h 11.6m, +28°43'*

Arp's Galaxy, *11h 19.6m, +51°30'*

Bode's "Nebulae" (M81 and M82, NGC 3031 and 3034), *9h 55.6m, +69°04' and 9h 55.8m, +69°41'*

Coddington's "Nebula" (IC 2574), *10h 28.4m, +68°25'*

Helix Galaxy (NGC 2685), *8h 55.6m, +58°44'*

Holmberg I, *9h 40.5m, +71°11'*

Holmberg II, *8h 18.9m, +70°43'*

Holmberg IV, *13h 54.8m, +53°54'*

Holmberg V, *13h 40.7m, +54°20'*

Holmberg IX, *9h 57.6m, +69°03'*

Keenan's System (NGC 5216 and 5218), *13h 32.1m, +62°42' and 13h 32.2m, +62°46'*

M81 (Bode's "Nebulae" with NGC 3034, NGC 3031), *9h 55.6m, +69°04'*

M81 Dwarf B, *10h 05.5m, +70°22'*

M81 Group, *10h 02.1m, +68°45'*

Ursa Major (cont.):

    galaxies (cont.):

        M82 (Bode's "Nebulae" with NGC 3031, NGC 3034,
            member of M81 group), *9h 55.8m, +69°41'*

        M101 (NGC 5457, Pinwheel Galaxy), *14h 03.2m, +54°21'*

        M101 Group, *14h 09.4m, +54°55'*

        M108 (NGC 3556), *11h 11.5m, +55°40'*

        M109 (NGC 3992), *11h 57.6m, +53°23'*

        Mayall's Object, *11h 03.9m, +40°51'*

        Miniature Spiral Galaxy (NGC 3928), *11h 51.8m, +48°41'*

        Papillon (IC 708), *11h 33.9m, +49°03'*

        Pinwheel Galaxy (M101, NGC 5457), *14h 03.2m, +54°21'*

        Shakhbazian 1 group, *10h 54.8m, +40°28'*

        Ursa Major I Cluster, *11h 47.0m, +55°44'*

        Ursa Major II Cluster, *10h 58.0m, +56°46'*

        Zwicky 2, *11h 58.4m, +38°04'*

    nebula, planetary, M97 (NGC 3587, Owl Nebula),
        11h 14.8m, +55°01'

    stars:

        alpha (α) Ursae Majoris (Dubhe), *11h 03.7m, +61°45'*

        alpha and beta (α and β) (Pointer Stars), *11h 03.7m,*
            +61°45' and 11h 01.8m, +56°23'

        beta (β) Ursae Majoris (Merak, Mirak), *11h 01.8m, +56°23'*

        delta (δ) Ursae Majoris (Megrez), *12h 15.4m, +57°02'*

        80 Ursae Majoris (Alcor, g Ursae Majoris, Saidak),
            13h 25.2m, +54°59'

        epsilon (ε) Ursae Majoris (Alioth), *12h 54.0m, +55°58'*

        eta (η) Ursae Majoris (Alcaid, Alkaid, Benetnasch),
            13h 47.5m, +49°19'

        g Ursae Majoris (Alcor, 80 Ursae Majoris, Saidak),
            13h 25.2m, +54°59'

        gamma (γ) Ursae Majoris (Phad, Phaed, Phecda),
            11h 53.8m, +53°42'

Ursa Major (cont.):

    stars (cont.):

        Groombridge 1618, *10h 11.4m, +49° 27'*

        Groombridge 1830 (Flying Star, Runaway Star), *11h 53.0m, +37° 43'*

        iota (ι) Ursae Majoris (Dnoces, Talita, Talitha), *8h 59.2m, +48° 02'*

        Lalande 21185, *11h 03.3m, +35° 58'*

        lambda (λ) Ursae Majoris (Tania Borealis), *10h 17.1m, +42° 55'*

        M40 double star, *12h 22.2m, +58° 05'*

        mu (μ) Ursae Majoris (Tania Australis), *10h 22.3m, +41° 30'*

        nu (ν) Ursae Majoris (Alula Borealis), *11h 18.5m, +33° 06'*

        omicron (o) Ursae Majoris (Muscida), *8h 30.3m, +60° 43'*

        Sidus Ludovicianum, *13h 25.0m, +54° 53'*

        xi (ξ) Ursae Majoris (Alula Australis, El Acola), *11h 18.2m, +31° 32'*

        zeta (ζ) Ursae Majoris (Mizar), *13h 23.9m, +54° 56'*

Ursa Major Moving (open) Cluster, *12h 03.0m, +58° 00'*

Ursa Major I Cluster of galaxies, *11h 47.0m, +55° 44'*

Ursa Major II Cluster of galaxies, *10h 58.0m, +56° 46'*

Ursa Minor (abbreviated **UMi**, Lesser Bear, Little Dipper):

    galaxies:

        Polarissima Borealis (NGC 3172), *11h 50.0m, +89° 07'*

        Ursa Minor Dwarf Galaxy, *15h 08.8m, +67° 12'*

    stars:

        alpha (α) Ursae Minoris (Cynosura, North Star, Polaris, Pole Star), *2h 31.8m, +89° 16'*

        beta (β) Ursae Minoris (Kochab), *14h 50.7m, +74° 09'*

        delta (δ) Ursae Minoris (Yildun), *17h 32.2m, +86° 35'*

        gamma (γ) Ursae Minoris (Pherkad), *15h 20.7m, +71° 50'*

Ursa Minor Dwarf Galaxy, *15h 08.8m, +67° 12'*

Vv

Van Biesbroeck's Star in Aquila, *19h 17.0m, +5°09'*

Van Maanen's Star in Pisces, *0h 49.2m, +5°23'*

Vega (alpha (α) Lyrae, Harp Star, Wega), *18h 36.9m, +38°47'*

Veil (bright) Nebula in Cygnus (Bridal Veil Nebula; Cirrus Nebula; Cygnus Loop; NGC 6960, 6992, and 6995), *20h 45.7m, +30°43'; 20h 56.4m, +31°43'; and 20h 57.1m, +31°13'*

Vela (abbreviated **Vel**, Sails of Argo Navis):

    cluster, open, omicron (o) Velorum Cluster (IC 2391, Velorum), *8h 40.2m, –53°04'*

    nebula, bright, Gum Nebula, *8h 30.0m, –45°00'*

    nebula, dark, Bok's Valentine, *8h 25.5m, –51°01'*

    nebula, planetary, Eight-burst Nebula (NGC 3132), *10h 07.0m, –40°26'*

    stars:

        delta & kappa (δ & κ) Velorum and epsilon & iota (ε & ι) Carinae (False Cross), *8h 44.7m, –54°42'; 9h 22.1m, –55°01'; 8h 22.5m, –59°31'; and 9h 17.1m, –59°17'*

        gamma^2 (γ^2) Velorum (Regor, Spectral Gem of the Southern Skies, Suhail), *8h 09.5m, –47°20'*

        lambda (λ) Velorum (Suhail), *9h 08.0m, –43°26'*

        N Velorum, *9h 31.2m, –57°02'*

Velorum open cluster in Vela (IC 2391, omicron (o) Velorum Cluster), *8h 40.2m, –53°04'*

Vindemiatrix (epsilon (ε) Virginis), *13h 02.2m, +10°58'*

Virgo (abbreviated **Vir**, Virgin):

    galaxies:

        Eyes (NGC 4435 and 4438), *12h 27.7m, +13°05'* and *12h 27.8m, +13°01'*

Virgo (cont.):

    galaxies (cont.):

        Gibson Reaves 8 (GR 8), *12h 58.7m, +14° 13'*

        Holmberg VII, *12h 34.7m, +6° 18'*

        M49 (NGC 4472), *12h 29.8m, +8° 00'*

        M58 (NGC 4579), *12h 37.7m, +11° 49'*

        M59 (NGC 4621), *12h 42.0m, +11° 39'*

        M60 (NGC 4649), *12h 43.7m, +11° 33'*

        M61 (NGC 4303), *12h 21.9m, +4° 28'*

        M84 (NGC 4374), *12h 25.1m, +12° 53'*

        M86 (NGC 4406), *12h 26.2m, +12° 57'*

        M87 (NGC 4486, Virgo A), *12h 30.8m, +12° 24'*

        M89 (NGC 4552), *12h 35.7m, +12° 33'*

        M90 (NGC 4569), *12h 36.8m, +13° 10'*

        M104 (Darklane Galaxy, NGC 4594, Sombrero Galaxy), *12h 40.0m, –11° 37'*

        Reinmuth 80 (NGC 4517A), *12h 32.5m, +0° 23'*

        Siamese Twins (NGC 4567 and 4568), *12h 36.5m, +11° 15'* and *12h 36.6m, +11° 14'*

        Virgo Cluster (Field of the Nebulae), *12h 30.0m, +12° 23'*

        Wild's Triplet, *11h 46.8m, –3° 51'*

    stars:

        alpha (α) Virginis (Azimech, Spica), *13h 25.2m, –11° 10'*

        beta (β) Virginis (Zavijava), *11h 50.7m, +1° 46'*

        epsilon (ε) Virginis (Vindemiatrix), *13h 02.2m, +10° 58'*

        eta (η) Virginis (Zaniah), *12h 19.9m, –0° 40'*

        gamma (γ) Virginis (Antevorta, Porrima), *12h 41.7m, –1° 27'*

        iota (ι) Virginis (Syrma), *14h 16.0m, –6° 00'*

        Olber's Star, *13h 14.1m, –16° 33'*

Virgo A Galaxy (M87, NGC 4486), *12h 30.8m, +12° 24'*

Virgo Cluster of galaxies (Field of the Nebulae), *12h 30.0m, +12° 23'*

Volans (abbreviated **Vol**, Flying Fish, shortened form of Piscis Volantis):

    galaxies:

        Das Rheingold (Graham's Object, Niebelungen Ring), *6h 41.4m, –74° 19'*

        Lindsay-Shapley Ring, *6h 43.1m, –74° 14'*

Vulpecula (abbreviated **Vul**, Fox, Little Fox):

    cluster, open, Brocchi's Cluster (Coat Hanger), *19h 25.4m, +20° 11'*

    nebula, planetary, M27 (Diabolo Nebula, Double-Headed Shot, Dumbbell Nebula, NGC 6853), *19h 59.6m, +22° 43'*

    star, Kuwano's Object, *20h 21.2m, +21° 36'*

Wachmann's Flare Star in Orion, *5h 33.8m, +1°57'*

Wagoner. *See* **Auriga**

Walborn's Star in Dorado, *4h 55.3m, –67°11'*

Warren and Penfold's (WP) Star in Dorado, *5h 39.0m, –64°05'*

Warrior. *See* **Orion**

Warrior King. *See* **Cepheus**

Wasat (delta (δ) Geminorum), *7h 20.1m, +21°59'*

Water Bearer. *See* **Aquarius**

Water Jar. *See* **Aquarius**

Water Monster. *See* **Hydra**

Water Snake (**Hydrus**, abbreviated **Hyi**)

Wazn:

    beta (β) Centauri (Agena, Hadar), *14h 03.8m, –60°22'*

    beta (β) Columbae (Wezn), *5h 51.0m, –35°46'*

Wega (alpha (α) Lyrae, Harp Star, Vega), *18h 36.9m, +38°47'*

Wesen (delta (δ) Canis Majoris, Wezen), *7h 08.4m, –26°24'*

Wezn (beta (β) Columbae, Wazn), *5h 51.0m, –35°46'*

Whale. *See* **Cetus**

Whirlpool Galaxy in Canes Venatici (Lord Rosse's "Nebula," M51, NGC 5194 and 5195, Question Mark Galaxy), *13h 29.9m, +47°12'* and *13h 30.0m, +47°16'*

Wild Duck (open) Cluster in Scutum (M11, NGC 6705), *18h 51.1m, –6°16'*

Wild's Triplet of galaxies in Virgo, *11h 46.8m, –3°51'*

Winged Horse. *See* **Pegasus**

Witch Head (bright) Nebula in Eridanus (IC 2118), *5h 06.9m, –7°13'*

WLM galaxy in Cetus (Wolf-Lundmark-Melotte System), *0h 02.0m, –15°28'*

Wolf (**Lupus**, abbreviated **Lup**)

Wolf-Lundmark-Melotte (WLM) System galaxy in Cetus, *0h 02.0m, –15°28'*

Woman Chained. *See* **Andromeda**

Wonderful Star (Mira, omicron (o) Ceti), *2h 19.3m, –2°59'*

WP (Warren and Penfold's) Star in Dorado, *5h 39.0m, –64°05'* xi (ξ) Cephei (Kurhah), *22h 03.8m, +64° 38'*

xi (ξ) Draconis (Grumium), *17h 53.5m, +56° 52'*

xi (ξ) Persei (Menkib), *3h 59.0m, +35° 47'*

xi (ξ) Ursae Majoris (Alula Australis, El Acola), *11h 18.2m, +31° 32'*

Yy

Y Canum Venaticorum (La Superba), *12h 45.1m, +45°26'*
Yed Posterior (epsilon (ε) Ophiuchi), *16h 18.3m, –4°42'*
Yed Prior (delta (δ) Ophiuchi), *16h 14.3m, –3°42'*
Yildun (delta (δ) Ursae Minoris), *17h 32.2m, +86°35'*

Zz

Zaniah (eta (η) Virginis), *12h 19.9m, –0°40'*

Zaurak (gamma (γ) Eridani), *3h 58.0m, –13°31'*

Zavijava (beta (β) Virginis), *11h 50.7m, +1°46'*

zeta (ζ) Aquilae (Deneb), *19h 05.4m, +13°52'*

zeta (ζ) Aurigae (Sadatoni), *5h 02.5m, +41°05'*

zeta (ζ) Cancri (Tegmen), *8h 12.2m, +17°39'*

zeta (ζ) Canis Majoris (Furud, Phurud), *6h 20.3m, –30°04'*

zeta (ζ) Ceti (Baten Kaitos), *1h 51.5m, –10°20'*

zeta (ζ) Geminorum (Mekbuda), *7h 04.1m, +20°34'*

zeta (ζ) Leonis (Adhafera, Aldhafera), *10h 16.7m, +23°25'*

zeta (ζ) Orionis (Alnitak), *5h 40.8m, –1°57'*

zeta (ζ) Pegasi (Homam), *22h 41.5m, +10°50'*

zeta (ζ) Persei (Atik), *3h 54.1m, +31°53'*

zeta (ζ) Puppis (Naos), *8h 03.6m, –40°00'*

zeta (ζ) Sagittarii (Ascella), *19h 02.6m, –29°53'*

zeta (ζ) Sculptoris (open) Cluster, *0h 04.3m, –29°56'*

zeta (ζ) Ursae Majoris (Mizar), *13h 23.9m, +54°56'*

Zosma (delta (δ) Leonis, Dhur, Duhr, Zubra), *11h 14.1m, +20°31'*

Zubenelgenubi (alpha² (α²) Librae, Kiffa Australis), *14h 50.9m, –16°02'*

Zubeneschamali (beta (β) Librae, Kiffa Borealis), *15h 17.0m, –9°23'*

Zubra (delta (δ) Leonis, Dhur, Duhr, Zosma), *11h 14.1m, +20°31'*

Zwicky 2 galaxy in Ursa Major, *11h 58.4m, +38°04'*

Zwicky's Triplet of galaxies in Hercules, *16h 49.5m, +45°30'*

About the Author

Hugh C. Maddocks first became interested in astronomy as a teenager in the fifties, using 4" (100 mm) reflector and 2.4" (60 mm) refractor telescopes for observations from his home in Milford, Connecticut and summer cottage near Lake Willoughby in northeastern Vermont. During the International Geophysical Year (IGY) in '57 and '58, he helped take data on meteor showers and northern lights.

In the sixties and early seventies, astronomy took a back seat to amateur radio and earning bachelor's and doctoral degrees in electrical engineering at the University of Vermont. He has practiced communications systems engineering for many years and has done professional technical indexing for major publishers since the mid-eighties.

When he quit smoking in the fall of '89, he decided that he needed an activity in addition to amateur radio to keep his hands busy, and his interest in observational astronomy returned. He now uses 14 x 100 giant binoculars with a homemade tripod mount and an 8" (200 mm) Schmidt-Cassegrain telescope to transform the light-polluted skies of northern Virginia into the dark skies of northeastern Vermont. His interests include rich star fields, clusters, and variable stars.

Notes

Notes

Notes

Notes

Notes

Feedback Form

(This feedback form may be photocopied.)

NAME ____

ADDRESS ____

____

____

____

PHONE ____

Additions and corrections to *Deep-Sky Name Index 2000.0* are welcomed from both amateur and professional astronomers. Please list each change below with at least one precise citation from a book, magazine, or professional journal which can be used to confirm the change.

MAIL TO:

Foxon-Maddocks Associates
10807 Oldfield Drive
Reston, Virginia 22091–5207 U.S.A.

Phone: (703) 476-4860

Order Form

(This order form may be photocopied.)

Please send the order below to:

NAME ____

ADDRESS ____

____

____

____

I would like to order:

_______ copies of *Deep-Sky Name Index 2000.0* at \$16.95 each
(postpaid in the U.S., U.S. possessions, and Canada)

Subtotal _____________

Virginia residents please add 4.5% sales tax _____________

Overseas orders please add \$5.00 per copy
for airmail shipment _____________

TOTAL Amount Enclosed (U.S. dollars only) _____________

Payment must be in U.S. dollars by check or bank draft on a
U.S. bank or by International Money Order.

MAIL TO:

Foxon-Maddocks Associates
10807 Oldfield Drive
Reston, Virginia 22091–5207 U.S.A.

Phone: (703) 476-4860